100% High School Mathematics Challenge

10 Practice Tests with Full Detailed Solutions

Sinan Kanbir, Ph.D.
Richard Spence, Ph.D(c).

November 2020

MathTopia Academy
ISBN: 978-1-7356252-4-9 (paper)
ISBN: 978-1-7356252-5-6 (ebook)

Acknowledgments

We would like to thank Dursun Caliskan for test solving and reading the drafts of this book and giving very valuable comments. We are also thankful to Kevin Schoenecker, University of Wisconsin System mathematics professor, for his willingness to proofread this book and always express his enjoyment. Last, but not least, we would like to thank University of Wisconsin-Stevens Point Mathematical Sciences department faculty members, Andy Felt, Hurlee Gonchigdanzan, and Dale M. Rohm for giving valuable feedback and always being willing to work problems in the hallways

Possible target audiences of this book

- Mathematics contest participants, such as AMC 10, AMC 12, MathCON, and Math Leagues.
- Math Circle students and organizers
- Pre-service and in-service secondary mathematics teachers.
- Parents of mathematically talented students.
- General math enthusiasts.

Practice Exams

Practice Exam 1

- You have **75 minutes** for **25 problems**.
- There are no penalties for incorrect answers. Answer as many problems as you can; return to the others in the time you have left for the test.

Problem 1. If $2^{2018} - 2^{2017} - 2^{2016} + 2^{2015} = a \times 2^{2015}$, then what is the value of a?

A) 3 B) 5 C) 8 D) 13 E) 16

Problem 2. How many digits are in the number $125^4 \times 64^2$, when written in base 10?

A) 10 B) 11 C) 12 D) 13 E) 14

Problem 3. A restaurant offers four appetizers, five entrées, and three desserts. Leah will order one entrée, at most one appetizer, and at most one dessert. How many different meal combinations can Leah order?

A) 12 B) 60 C) 75 D) 90 E) 100

Problem 4. How many of the following five shapes could be the shape of the region where two triangles overlap?

I. *equilateral triangle* II. *regular pentagon* III. *regular hexagon*

IV. *square* V. *kite*

A) 1 B) 2 C) 3 D) 4 E) 5

Problem 5. Five students stand in line from shortest to tallest. The average height of the three shortest students is 58 inches, and the average height of the three tallest students is 70 inches. The average height of all five students is 63 inches. What is the median height of the five students, in inches?

A) 59 B) 64 C) 67 D) 68 E) 69

Problem 6. For how many integers n is $|n^2 - 6n + 5|$ prime?

A) 1 B) 2 C) 3 D) 4 E) 5

Problem 7. Maria flips six fair coins. What is the probability that at least half of the coins land heads?

A) $\frac{10}{32}$ B) $\frac{1}{2}$ C) $\frac{41}{64}$ D) $\frac{21}{32}$ E) None of the preceding

Problem 8. A rectangular box has integer edge lengths in the ratio $1 : \frac{3}{2} : 2$. Which of the following could be the volume of the box?

A) 120 B) 144 C) 168 D) 192 E) 216

Problem 9. The sum of the first 20 terms of an arithmetic sequence which has first term 1 is equal to the sum of the first 10 terms of an arithmetic sequence which has first term 10. If positive integers x and y are the respective common differences of the sequences, what is the minimum possible value of $x + y$?

A) 35 B) 38 C) 40 D) 43 E) None of the preceding

Problem 10. The number $N = 11111 \ldots 111$ is formed by writing 105 ones in a row. What is the sum of the digits of the product $105 \times N$?

A) 504 B) 555 C) 630 D) 945 E) 1105

Problem 11. Mark writes the number 1 on a blackboard. Every second, Mark flips a fair coin. If the coin lands heads, Mark replaces the number n on the blackboard with $2n$; if the coin lands tails, he replaces the number with $\frac{n}{2}$. After 10 seconds, what is the expected value of the number on the blackboard?

A) 1 B) $\frac{5}{4}$ C) $\frac{3}{2}$ D) $\frac{5^{10}}{4^{10}}$ E) $\frac{3^{10}}{2^{10}}$

Problem 12. How many non-congruent isosceles triangles can be created with integer side lengths and a perimeter of 200?

A) 49 B) 50 C) 51 D) 64 E) 65

Problem 13. The operation $a \spadesuit b$ is defined by $a \spadesuit b = ab - 2(a+b)$ for all integers a and b. What is the value of $1 \spadesuit (2 \spadesuit (3 \spadesuit (4 \spadesuit (5 \spadesuit 6))))$?

A) -10 B) -4 C) -2 D) 2 E) 10

Problem 14. Given a positive integer n, let $d(n)$ denote the number of positive divisors of n. For example, $d(6) = 4$ and $d(17) = 2$. Let m be the smallest positive integer such that $d(m) = d(m+1) = 4$. What is the sum of the digits of m?

A) 3 B) 5 C) 8 D) 9 E) 10

Problem 15. Five soccer teams participate in a tournament. Each team plays all the other teams exactly once. The winning team of each game is awarded 3 points and the losing team is awarded no points. If a game ends in a tie, both teams get 1 point. At the end of the tournament, four of the five teams were awarded 1, 2, 5, and 8 points in total. How many points were awarded to the fifth team?

A) 10 B) 12 C) 14 D) 15 E) None of the preceding

Problem 16. How many non-congruent quadrilaterals $ABCD$ can be constructed such that $\angle DAB = 30°$, $\angle BCD = 90°$, $AB = 17$, $BC = 13$ and $CD = 11$?

A) 0 B) 1 C) 2 D) 3 E) Infinitely many

Problem 17. A parabola with equation $y = ax^2 + bx + c$ has vertex (h, k). How many of the six quantities a, b, c, h, k and $\Delta = b^2 - 4ac$ can be negative at the same time?

A) At most 2 B) At most 3 C) At most 4 D) At most 5 E) All six

Problem 18. Suppose $x < y < z$ are prime numbers such that $x+y+z = 68$ and $xy+xz+yz = 437$. What is the value of yz?

A) 245 B) 295 C) 305 D) 371 E) 427

Problem 19. How many positive integers less than or equal to 2^{10} are there whose binary representation contains two or more consecutive 1's? One example is 27, since $27 = 11011_2$.

A) 879 B) 880 C) 934 D) 968 E) 969

Problem 20. In triangle ABC, medians $\overline{BE}$ and $\overline{CD}$ intersect at G and are perpendicular to each other. If $BE = 18$ and $CD = 24$, what is the length of $\overline{AG}$?

A) 15 B) 20 C) 25 D) 30 E) None of the preceding

Problem 21. Suppose a and b are real numbers such that $ab^2 = 1$ and $a^3 + 3b^3 = 4$. What is the product of all possible values of $a^3 + b^3$?

A) 12 B) 18 C) 24 D) 36 E) 72

Problem 22. The sum of two positive integers a and b is 2020. The number a has exactly nine positive divisors and the number b has exactly 21 positive divisors. What is $a - b$?

A) -1508 B) -868 C) 868 D) 1508 E) None of the preceding

Problem 23. For all integers $i \geq 0$, let P_i denote the point $\left(2^i, \frac{2020}{2^i}\right)$ in the xy-plane. What is the smallest integer n such that the area of convex polygon $P_0P_1P_2 \ldots P_n$ is greater than 2020^2?

A) 10 B) 11 C) 12 D) 13 E) 16

Problem 24. The real number $\sqrt[3]{6\sqrt{3}+10}-\sqrt[3]{6\sqrt{3}-10}$ is a root of the cubic polynomial x^3+bx+c for some integers b and c. What is $b+c$?

A) -14 B) -8 C) 8 D) 14 E) Cannot be determined

Problem 25. Six teams numbered #1, #2, ..., #6 play a round robin tournament, where each team plays every other team exactly once in a match. For all $i \neq j$ with $1 \leq i,j \leq 6$, the probability that team #i beats team #j in their match is $\frac{1}{2}$, independently of other matches. What is the probability that every team wins at least one match and loses at least one match?

A) $\frac{5}{8}$ B) $\frac{335}{512}$ C) $\frac{169}{256}$ D) $\frac{175}{256}$ E) $\frac{89}{128}$

Practice Exam 2

- You have **75 minutes** for **25 problems**.
- There are no penalties for incorrect answers. Answer as many problems as you can; return to the others in the time you have left for the test.

Problem 1. Eight chess players play in a round-robin tournament, where each player plays every other player exactly once. A win is worth 1 point for the winner and 0 points for the loser, and a tie is worth 0.5 points for both players. Given that seven of the players' cumulative scores were 0, 1, 2.5, 3, 3.5, 5, and 7 points, how many points did the eighth player score?

A) 3 B) 4 C) 5 D) 6 E) 7

Problem 2. In the following multiplication problem, each of the letters A, B, and C represents a different digit between 1 and 9, inclusive. What is A?

$$\begin{array}{cccc} & & \mathrm{A} & \mathrm{B} \\ \times & & \mathrm{B} & \mathrm{C} \\ \hline \mathrm{C} & \mathrm{A} & \mathrm{C} & 1 \end{array}$$

A) 2 B) 4 C) 5 D) 8 E) 9

Problem 3. How many 4-digit positive integers containing only odd digits are divisible by 5?

A) 100 B) 125 C) 250 D) 500 E) 625

Problem 4. Rhombus $ABCD$ has $AB = 6$ and $\angle ABC = 60^\circ$. Point P is chosen in order to minimize the sum of the distances from P to each of the points A, B, C, and D. What is $PA + PB + PC + PD$?

A) $3 + 3\sqrt{3}$ B) 6 C) $6 + 3\sqrt{3}$ D) $6 + 6\sqrt{3}$ E) 18

Problem 5. Richard's average bowling score in a league is 180. Today, he bowled three league games and scored 193, 207, and 200, which raised his average to 182. Including the games bowled today, how many league games has Richard bowled thus far?

A) 24 B) 27 C) 30 D) 33 E) 36

Problem 6. Alice writes down a list of whole numbers beginning with 1. To generate the next number in the list, she either adds 6 to the previous number, or she multiplies the previous number by 4. For example, her list could be the sequence 1, 7, 28, 34, 40, 160, ...Which of the following numbers *cannot* appear in Alice's list?

A) 109 B) 151 C) 244 D) 335 E) 412

Problem 7. Robert has a coin which lands heads with probability $\frac{2}{3}$ and tails with probability $\frac{1}{3}$. He flips this coin three times. What is the probability that he obtains at least one tail?

A) $\frac{4}{9}$ B) $\frac{17}{27}$ C) $\frac{19}{27}$ D) $\frac{23}{27}$ E) $\frac{26}{27}$

Problem 8. Square ABCD has side length 4. Points E, F, G, and H are the midpoints of $\overline{\mathrm{AB}}$, $\overline{\mathrm{BC}}$, $\overline{\mathrm{CD}}$, and $\overline{\mathrm{DA}}$ respectively. What is the area of the shaded square?

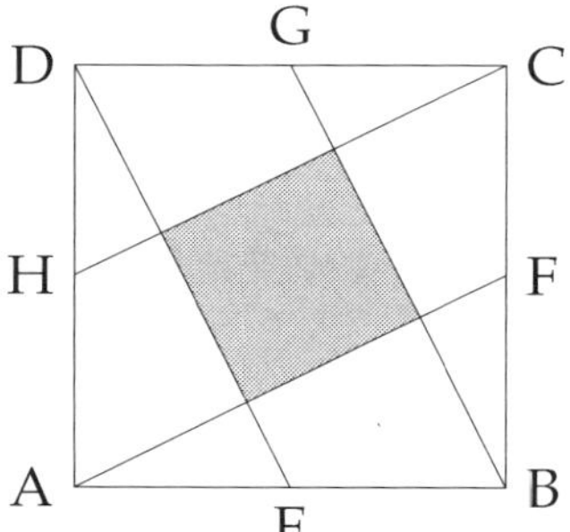

A) 2.4

B) 3.2

C) 3.6

D) 4

E) None of the preceding

Problem 9. The cubic polynomial $x^3 + 9x^2 + 24x + 16$ has exactly two distinct real roots, both of which are integers. What is the sum of these two roots?

A) -9 B) -5 C) 2 D) 5 E) 17

Problem 10. The last two digits of n (when written in base 10) are 99, and n has exactly six positive divisors. How many divisors does $100n$ have?

A) 24 B) 30 C) 54 D) 600 E) The answer depends on n.

Problem 11. James is setting up a 4-digit personal identification number (PIN). His PIN will consist of four *distinct* digits $\overline{abcd}$, where a and b are adjacent on the keypad, b and c are adjacent, and c and d are adjacent. For example, 1236, 0896, and 4587 are valid PINs, while 1563, 1232, and 1455 are not. How many possible 4-digit PINs can James create?

A) 82
B) 86
C) 88
D) 90
E) 92

Problem 12. Triangle ABC has $AB = 5$, $AC = 12$, and $m\angle A = 90^\circ$. Point D is on $\overline{AC}$ such that $\overline{BD}$ bisects $\angle ABC$, and point M is the midpoint of $\overline{BD}$. What is $\sin m\angle DAM$?

A) $\frac{2\sqrt{13}}{13}$ B) $\frac{2}{3}$ C) $\frac{4}{5}$ D) $\frac{3\sqrt{13}}{13}$ E) $\frac{12}{13}$

Problem 13. If $2^{2^{2020}} = 4^k$, then what is the value of k?

A) 2020 B) 2^{1010} C) 2^{2019} D) $2^{2020} - 1$ E) 2^{2020}

Problem 14. The increasing sequence 2, 4, 6, 8, 20, 22, 24, …contains all positive integers whose digits are even. The kth number in this sequence is 2020. What is k?

A) 125 B) 130 C) 135 D) 250 E) 260

Problem 15. Alice and Bob send each other messages using the encoding scheme A = 1, B = 2, C = 3, ..., Z = 26. For example, the four-letter string MATH is encoded as 13-1-20-8. Alice sent Bob an encoded sequence of numbers, but accidentally left out the dashes, so Bob received the sequence 131208919123519151 35. How many possible strings of letters could Alice have encoded? Two possible strings are MATHISAWESOME and ACATHISLCESAEACE.

A) 28 B) 48 C) 64 D) 96 E) 144

Problem 16. Isosceles trapezoid $ABCD$ has side lengths $AB = 10$, and $BC = CD = DA = 6$. Let ω be the circumcircle of trapezoid $ABCD$. What is the area of ω?

A) 25π B) 27π C) 30π D) 33π E) 36π

Problem 17. Let $f : A \to \mathbb{R}$ be defined by $f(x) = |\ln|\tan x||$, where $A = \mathbb{R}\ \{k\pi/2 : k \in \mathbb{Z}\}$ is the set of all real numbers which are not integer multiples of $\frac{\pi}{2}$. For how many values $x \in [0, 2020\pi]$ is it true that $f(x) = 2020$?

A) 1 B) 1010 C) 2020 D) 4040 E) 8080

Problem 18. Ayu is thinking of a positive five-digit integer. The number is divisible by 792, and it contains the digits 1, 2, 3, and 4. What is the hundreds digit of Ayu's number?

A) 1 B) 2 C) 3 D) 4 E) 8

Problem 19. A cashier has four \$1 bills and an unlimited supply of \$2 and \$5 bills. How many different ways can the cashier make change for \$100? For example, the cashier may use 20 \$5 bills, or 10 \$5 bills, 23 \$2 bills, and four \$1 bills.

A) 47 B) 48 C) 50 D) 51 E) 52

Problem 20. A right circular cone has a base radius of 3 cm and a height of 4 cm. A sphere is inscribed inside the cone, tangent to the base and the lateral surface of the cone. What is the volume of this sphere, in cm^3?

A) 3π B) $\frac{256}{81}\pi$ C) $\frac{27}{8}\pi$ D) 4π E) $\frac{9}{2}\pi$

Problem 21. Given that $x > 0$, what is the minimum possible value of the expression $x + \frac{1}{x} + 2x^2 + \frac{2}{x^2}$?

A) 1 B) 3 C) 4 D) 6 E) None of the preceding

Problem 22. Let $f^1(x) = x^2 - 20$ for all real numbers x, and let $f^n(x) = f^1(f^{n-1}(x))$ for all integers $n \geq 2$. Let x_0 and x_1 be the smallest and largest real solutions to the equation $f^{2020}(x) = 0$ respectively. What is the greatest integer less than or equal to $x_0^2 + x_1^2$?

A) 32 B) 40 C) 41 D) 49 E) 50

Problem 23. Richard flips a fair coin ten times. What is the probability that he never obtains three consecutive heads or three consecutive tails?

A) $\frac{55}{512}$ B) $\frac{117}{1024}$ C) $\frac{123}{1024}$ D) $\frac{79}{512}$ E) $\frac{89}{512}$

Problem 24. The polynomial $P(x) = x^3 + 2x + 2020$ has three distinct complex roots z_1, z_2, and z_3. The polynomial $Q(x) = x^3 + ax^2 + bx + c$ has roots z_1^2, z_2^2, and z_3^2. What is $a + b + c$?

A) -4080392 B) -4080391 C) -4080390 D) 4080390 E) 4080392

Problem 25. A mathematician offers you the following game. He will flip 10 fair coins, and if n coins land heads, then you win n^2 dollars (for example, if 4 coins land heads, then you win \$16, and if 0 coins land heads, then you win \$0). How many dollars should the mathematician charge to play the game so that the game is fair?
Note: A game is *fair* if the expected profit after playing the game is zero.

A) \$25 B) \$27.50 C) \$28 D) \$28.50 E) \$30

Practice Exam 3

- You have **75 minutes** for **25 problems**.
- There are no penalties for incorrect answers. Answer as many problems as you can; return to the others in the time you have left for the test.

Problem 1. Jack averaged 40 miles per gallon while driving from San Diego to Los Angeles. On the return trip, he took a detour which covered 50% more distance, and only averaged 20 miles per gallon due to slower traffic. Which of the following is closest to Jack's overall fuel efficiency (in miles per gallon) for the entire round trip?

A) 25.0 B) 26.7 C) 27.5 D) 28.0 E) 30.0

Problem 2. Lisa writes a sequence of numbers on a blackboard, beginning with 1. Each number after the first number is equal to the sum of the squares of the numbers already on the blackboard. The first five numbers of Lisa's sequence are 1, 1, 2, 6, and 42. What are the last two digits of the tenth number of Lisa's sequence?

A) 06 B) 26 C) 42 D) 46 E) 62

Problem 3. Carl has a spinner with equally-likely sectors labeled 1, 2, ..., 5. If Carl spins the spinner three times, what is the probability that the sum of the three numbers he obtains is even?

A) $\frac{61}{125}$ B) $\frac{62}{125}$ C) $\frac{1}{2}$ D) $\frac{63}{125}$ E) $\frac{64}{125}$

Problem 4. Rectangle $ABCD$ has $AB = 4$ and $BC = 3$. Points P and Q are on diagonal $\overline{BD}$ such that $\overline{AP}$ bisects $\angle A$, and $\overline{CQ}$ bisects $\angle C$. What is the area of quadrilateral $APCQ$?

A) $\frac{12}{7}$ B) 2 C) $\frac{12}{5}$ D) 3 E) None of the preceding

Problem 5. If $(n + 3^{2019})^2 - (n - 3^{2019})^2 = 3^{2019} + 3^{2020}$, then what is the value of n?

A) 1 B) 9 C) 3^{2019} D) 3^{2020} E) 3^{2038}

Problem 6. For how many integers n is the expression $\frac{n^2}{n+4}$ equal to an integer?

A) 4 B) 5 C) 8 D) 10 E) 12

Problem 7. Mr. V's school has a MathCON math team containing six boys and four girls. In how many distinguishable ways can they be divided into two teams of five with at least one girl on each of the teams? The order of the students on each team does not matter.

A) 60 B) 114 C) 120 D) 126 E) 252

Problem 8. A box is in the shape of a rectangular prism with length 12 in, width 3 in, and height 4 in. An ant crawls along the outer surface from one vertex to the opposite vertex of the box. Which of the following is closest to the minimum possible distance the ant must crawl?

A) 13.0 in B) 13.4 in C) 13.9 in D) 15.5 in E) 19.0 in

Problem 9. Let A, B, and C be nonzero digits so that the sum of the two-digit numbers $\overline{AB}$, $\overline{BC}$, and $\overline{CA}$ is equal to the three-digit number $\overline{ABC}$. What is the value of $A + B + C$?

A) 9 B) 15 C) 18 D) 19 E) Cannot be determined

Problem 10. If a, b and c are distinct prime numbers such that $a - c = 8300$ and $a + b + c = 9704$, then what is the value of $a + 2b$?

A) 705 B) 9001 C) 9005 D) 9702 E) None of the preceding

Problem 11. Three people seat themselves in a row of ten chairs so that there is at least one seat between any two people. In how many different ways can the three people seat themselves?

A) 56 B) 120 C) 252 D) 336 E) 720

Problem 12. The sides of a triangle have lengths 7, 14, and c, where c is an integer. For how many values of c is the triangle acute?

A) 2 B) 3 C) 4 D) 5 E) 8

Problem 13. If $\frac{(x-2020)\cdot(y-2019)}{(x-2020)^2+(y-2019)^2}=\frac{1}{2}$, what is the value of $x-y$?

A) -1 B) 1 C) 2019 D) 2020 E) 4039

Problem 14. Given a positive integer n, let $s(n)$ denote the sum of the digits of n. For example, $s(5081)=5+0+8+1=14$. How many positive integers n satisfy $s(n)+3n=2019$?

A) 0 B) 1 C) 2 D) 3 E) 4

Problem 15. A MathCON exam consists of 32 problems, where each problem has five choices labeled A, ..., E, only one of which is correct. Problems 1 through 8 are worth 5 points each, problems 9 through 24 are worth 7 points each, and problems 25 through 32 are worth 8 points each. Incorrect or unanswered problems are worth 0 points, and the maximum score is 216. If Max randomly guesses one answer choice on every problem, then the probability that Max scores exactly 200 points is $\frac{m}{5^{32}}$ for some integer m. What is m?

A) 28 B) 120 C) 336 D) 448 E) 700

Problem 16. The figure below contains 25 points arranged in a 5×5 grid, where each pair of adjacent points is connected by a segment of length 1. What is the largest possible number of segments that can be removed from the figure so that it is still possible to travel from any point to any other point via a path of length at most 8?

A) 12
B) 14
C) 15
D) 16
E) 18

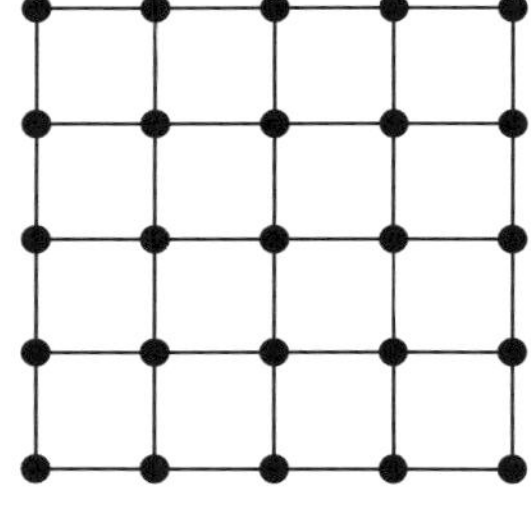

Problem 17. The complex numbers r_1, r_2, and r_3 are the roots of the polynomial $x^3 + 2020x - 1$. What is the value of $r_1^2 + r_2^2 + r_3^2$?

A) -2020^2 B) -4040 C) 0 D) 4040 E) 2020^2

Problem 18. Let $N = 12345678910111213\ldots394041$ be the 73-digit number obtained that is formed by writing the integers from 1 to 41 in order, one after the other. What is the remainder when N is divided by 9?

A) 0 B) 3 C) 6 D) 7 E) 8

Problem 19. Each of the nine circles is to contain one of the numbers 1, 2, or 3 such that the sum of the four numbers in each of the four unit squares is divisible by 3. In how many ways can this be done? Two ways that differ only by rotation or reflection are considered distinct.

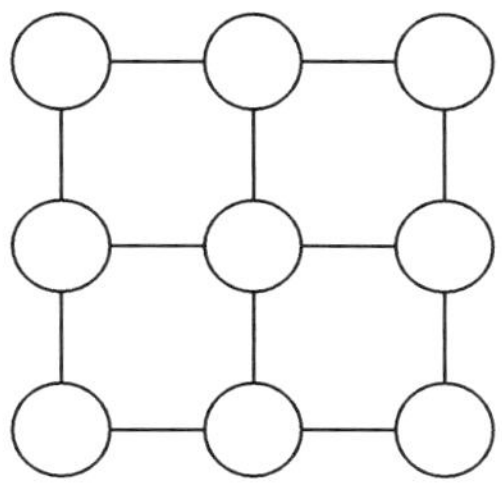

A) 3^5 B) 3^6 C) 3^7 D) 3^8 E) 3^9

Problem 20. In triangle ABC, point M is on $\overline{AC}$ and point N is on $\overline{BC}$. Line segments $\overline{BM}$ and $\overline{AN}$ intersect at point O. Given that the areas of $\triangle OMA$, $\triangle OBN$, and $\triangle OAB$ are 21, 42, and 63, respectively, what is the area of $\triangle MNC$?

A) 40 B) 45 C) 49 D) 56 E) 63

Problem 21. The sum of two real numbers is equal to a positive integer n, and the sum of their squares is $n + 19$. What is the maximum possible value of n?

A) 5 B) 7 C) 8 D) 9 E) 11

Problem 22. Point O is inside equilateral triangle ABC. Lines ℓ_1, ℓ_2, and ℓ_3 are drawn through O, parallel to sides AB, BC, and AC, respectively. These three lines divide $\triangle ABC$ into six non-overlapping polygons, three of which are equilateral triangles and three of which are parallelograms. If the areas of the three parallelograms are $48\sqrt{3}$, $60\sqrt{3}$, and $80\sqrt{3}$, then what is the area of $\triangle ABC$?

A) $216\sqrt{3}$ B) $240\sqrt{3}$ C) $282\sqrt{3}$ D) $288\sqrt{3}$ E) None of the preceding

Problem 23. How many of the first 1000 positive integers can be written in the form

$$\left\lfloor \frac{n}{3} \right\rfloor + \left\lfloor \frac{n}{4} \right\rfloor + \left\lfloor \frac{n}{5} \right\rfloor$$

for some positive integer n? The notation $\lfloor x \rfloor$ denotes the largest integer less than or equal to x.

A) 746 B) 767 C) 788 D) 796 E) 809

Problem 24. Two distinct lines ℓ_1 and ℓ_2 in the xy-plane pass through the origin and are tangent to the circle $(x-2)^2+(y-3)^2=1$. What is the product of the slopes of ℓ_1 and ℓ_2?

A) 2 B) $\frac{9}{4}$ C) $\frac{8}{3}$ D) $\frac{12}{5}$ E) $\frac{10}{3}$

Problem 25. A positive integer $n \geq 2$ is called *good* if there exists an integer $2 \leq \ell \leq 12$ and a length ℓ sequence $a_1, a_2, \dots, a_\ell$ satisfying the following properties:

- $a_1 = 1$
- $a_\ell = n$
- For all $2 \leq i \leq \ell$, either $a_i = a_{i-1} + 1$ or $a_i = 2a_{i-1}$

For example, 28 is good since there exists a length $\ell = 7$ sequence satisfying the above properties: 1, 2, 3, 6, 7, 14, 28. How many positive integers are good?

A) 374 B) 375 C) 376 D) 377 E) 378

Practice Exam 4

- You have **75 minutes** for **25 problems**.
- There are no penalties for incorrect answers. Answer as many problems as you can; return to the others in the time you have left for the test.

Problem 1. The arithmetic mean of x and y is 12, the arithmetic mean of x and z is 6, and the arithmetic mean of y and z is 4. Which of the following is equal to z?

A) -5 B) -2 C) 2 D) 10 E) 20

Problem 2. A positive integer $n \geq 2$ is *almost prime* if it is composite, but not divisible by 2, 3, 5, or 7. How many positive integers less than 300 are almost prime?

A) 6 B) 7 C) 8 D) 9 E) 10

Problem 3. Two cards are randomly dealt from a standard 52-card deck, without replacement. What is the probability that the two cards are of the same suit?
Note: A standard 52-card deck contains four suits, with 13 cards from each suit.

A) $\frac{1}{17}$ B) $\frac{3}{17}$ C) $\frac{3}{13}$ D) $\frac{4}{17}$ E) $\frac{1}{4}$

Problem 4. A very large floor is tessellated with octagon and square tiles, as shown below:

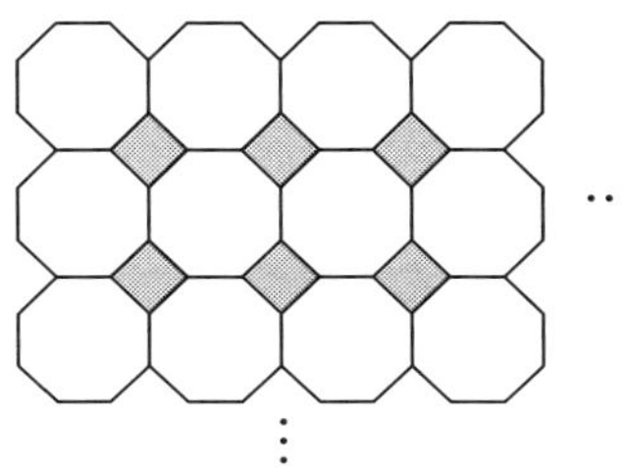

Approximately what percent of the area of the floor is covered with square tiles?

A) 14% B) 17% C) 21% D) 25% E) 28%

Problem 5. How many ordered triples of positive integers (a, b, c) with $a \neq b$ satisfy the equation $a! \times b! = c^2$?

A) 1 B) 2 C) 3 D) More than 3, but finitely many E) Infinitely many

Problem 6. How many positive divisors of 20^{20} are not divisible by 20?

A) 45 B) 61 C) 79 D) 81 E) 120

Problem 7. Let $S = \{1, 2, 3, \dots, 30\}$. How many subsets of S of size 4 are there whose elements are in arithmetic progression? One example is $\{12, 14, 16, 18\}$.

A) 120 B) 126 C) 129 D) 132 E) 135

Problem 8. A sphere is inscribed inside a cube of side length 2 cm. A smaller sphere is tangent to three of the faces of the cube as well as the larger sphere. What is the radius of the smaller sphere?

A) $5 - 3\sqrt{3}$ B) $\frac{1}{4}$ C) $2 - \sqrt{3}$ D) $\frac{3\sqrt{3} - 4}{4}$ E) $\frac{\sqrt{3} - 1}{2}$

Problem 9. The quadratic equation $x^2 + 8x - 84 = 0$ has two real solutions. If these solutions are plotted on the real number line, what is the distance in units between these points?

A) 8 B) 10 C) 15 D) 20 E) 28

Problem 10. What is the remainder when 5^{100} is divided by 72?

A) 1 B) 23 C) 49 D) 53 E) 71

Problem 11. A cashier has an unlimited supply of pennies, nickels, dimes, and quarters. In how many ways can the cashier give exactly 47 cents to a customer?

A) 21 B) 27 C) 33 D) 39 E) 47

Problem 12. If $\tan\alpha + \cot\alpha = \frac{13}{6}$, where $\alpha \in \left(0, \frac{\pi}{4}\right)$, what is the value of $\sin\alpha$?

A) $\frac{5}{13}$ B) $\frac{2\sqrt{13}}{13}$ C) $\frac{2}{3}$ D) $\frac{3\sqrt{13}}{13}$ E) $\frac{12}{13}$

Problem 13. If $\log_2 3 = a$ and $\log_2 5 = b$, then which of the following is equal to $\log_{225} 256$?

A) $\frac{2}{a+b}$ B) $\frac{2}{a^2+b^2}$ C) $\frac{4}{a+b}$ D) $\frac{4}{a^2+b^2}$ E) $\frac{8}{a+b}$

Problem 14. How many positive three-digit integers leave *different* remainders when divided by 9 and 10?

A) 630 B) 675 C) 720 D) 765 E) 810

Problem 15. Richard rolls a fair six-sided die; let n be the number showing on the top face. He then rolls n fair six-sided dice, and adds the n numbers showing up on the top faces. What is the *expected* value of Richard's sum?

A) 6 B) $\frac{7}{2}$ C) 12 D) $\frac{49}{4}$ E) 36

Problem 16. The quarter circle shown has radius 6. Rectangle $OABC$ has perimeter 16. If the perimeter of the shaded region is $3\pi + m$, what is the value of m?

A) 6
B) $6\sqrt{2}$
C) $8\sqrt{2}$
D) 10
E) 12

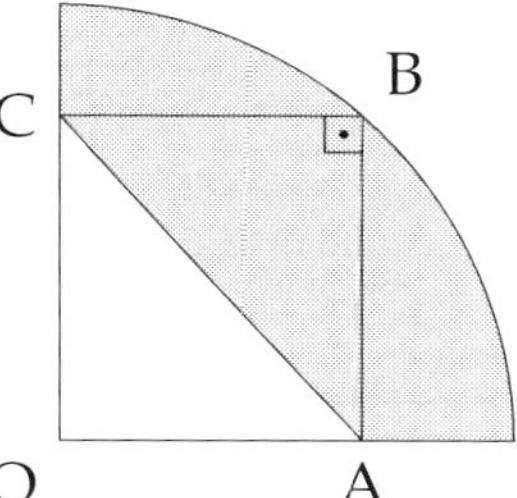

Problem 17. Given that $0 \le x \le 1$, what is the maximum possible value of $x\sqrt{1-x^2}$?

A) $\frac{1}{4}$ B) $\frac{\sqrt{3}}{4}$ C) $\frac{1}{2}$ D) 1 E) The maximum does not exist.

Problem 18. How many positive divisors of 10^{10} leave a remainder of 1 when divided by 3?

A) 50 B) 60 C) 61 D) 100 E) 121

Problem 19. How many ordered pairs (x, y) of integers satisfy the equation $(y + x)(y - x) = 111 + 6y$?

A) 1 B) 2 C) 4 D) 8 E) 16

Problem 20. Triangle ABC has $AB = 2$ and $AC = 4$. Point D is on $\overline{BC}$ such that $AD = 2BD$ and $DC = 3BD$. If R_1 and R_2 are the radii of the circumscribed circles of $\triangle ABD$ and $\triangle ADC$, respectively, what is $\frac{R_1}{R_2}$?

A) $\frac{1}{4}$ B) $\frac{1}{2}$ C) $\frac{3}{5}$ D) $\frac{3}{4}$ E) None of the preceding

Problem 21. Let a, b, and c be non-negative real numbers. Given that $a + 2b + 3c = 1$, what is the smallest possible value of $a^2 + b^2 + c^2$?

A) $\frac{1}{28}$ B) $\frac{1}{14}$ C) $\frac{1}{9}$ D) $\frac{49}{324}$ E) None of the preceding

Problem 22. In the semicircle shown, $\overline{AB}$ is a diameter with $AB = 8$. Points C and D are on the semicircle such that $BC = CD = 2$. What is the length of $\overline{AD}$?

A) $4\sqrt{3}$

B) 7

C) $2\sqrt{14}$

D) $2\sqrt{15}$

E) 8

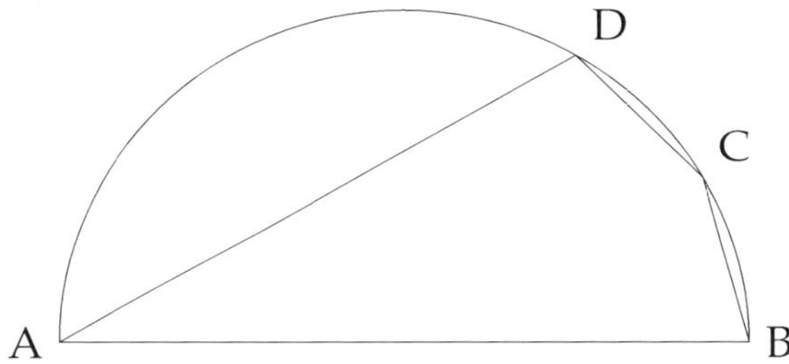

Problem 23. The increasing sequence 1, 3, 9, 10, 27, 28, 30, ... consists of all positive integers which can be written as a sum of one or more distinct, non-consecutive powers of 3. What is the 100th term of this sequence?

A) 19767 B) 19773 C) 19774 D) 19926 E) 19927

Problem 24. How many increasing sequences of positive integers are there which satisfy the following properties?

- The first term is 1.
- The last term is 20.
- The positive difference between any two consecutive terms is at least 3.

A) 129 B) 156 C) 189 D) 243 E) 277

Problem 25. The polynomial $x^3 - 5x + 1$ has three distinct real roots r, s, and t. What is the value of $\frac{r^2}{s} + \frac{s^2}{t} + \frac{t^2}{r}$?

A) -25 B) -5 C) 5 D) 25 E) 125

Practice Exam 5

- ♦ You have **75 minutes** for **25 problems**.
- ♦ There are no penalties for incorrect answers. Answer as many problems as you can; return to the others in the time you have left for the test.

Problem 1. A grocery store sells bags of tortilla chips for $2.50 each. However, there is a sale for 30% off each bag if five or more bags of tortilla chips are bought. Excluding tax, how many dollars does Victor save if he buys five bags of tortilla chips instead of four?

A) $1.00 B) $1.25 C) $1.75 D) $2.25 E) $3.75

Problem 2. How many ordered pairs of integers (a, b) satisfy the equation $a^2 - b^2 = 2020$?

A) 1 B) 2 C) 4 D) 8 E) 16

Problem 3. Four squares are drawn in the same plane. There are n points for which two or more of these squares intersect. What is the largest possible value of n?

A) 16 B) 32 C) 36 D) 48 E) 64

Problem 4. Given that $ABCDEFGH$ is a regular octagon, what is the degree measure of $\angle AGH$?

A) 15° B) 22.5° C) 25° D) 27.5° E) 30°

Problem 5. How many integers x are solutions to the equation $(x^2 - 3x + 1)^{x+1} = 1$?

A) 1 B) 2 C) 3 D) 4 E) 5

Problem 6. A *deficient* number is a positive integer n such that the sum of the proper divisors of n is less than n. For example, 10 is a deficient number since $1+2+5=8<10$. If n is a deficient number, which of the following numbers must also be deficient? (Note: $\lfloor x \rfloor$ denotes the largest integer less than or equal to x.)

A) $\lfloor \frac{n}{2} \rfloor$ B) $2n$ C) n^2-1 D) n^2 E) None of the preceding

Problem 7. Five companies each send three representatives to a networking event. At the event, each representative shakes representatives with all representatives from companies other than his or her own. How many handshakes take place?

A) 72 B) 90 C) 108 D) 120 E) 156

Problem 8. In the coordinate plane shown below, $\overline{YX} \perp \overline{YZ}$. What is the value of z?

A) 5
B) 6
C) 7
D) 8
E) 9

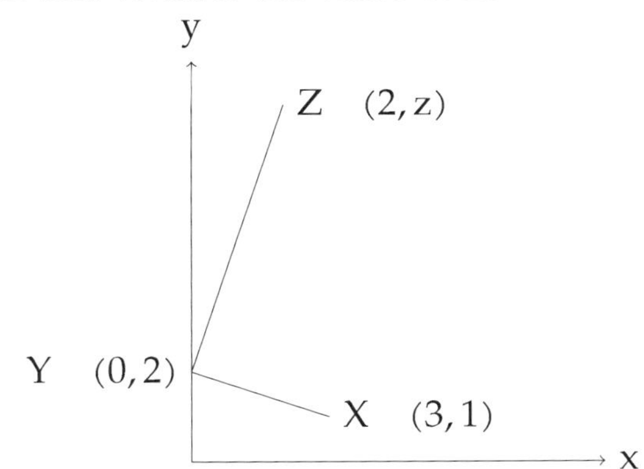

Problem 9. The second term of a geometric sequence is 1 more than the first term, and the third term of the same sequence is 4 more than the second term. What is the fourth term of this sequence?

A) $\frac{4}{3}$ B) $\frac{8}{3}$ C) 5 D) $\frac{16}{3}$ E) $\frac{64}{3}$

Problem 10. The pages of a book are numbered starting with 1. The sum of the page numbers in the book is less than 2020. If there was one more page, then the sum of the page numbers in the book would be greater than 2020. How many pages does the book have?

A) 60 B) 61 C) 62 D) 63 E) 64

Problem 11. How many three-element subsets of $\{1, 2, 3, \dots, 12\}$ are there whose sum is divisible by 3? One example is $\{1, 2, 9\}$.

A) 64 B) 68 C) 72 D) 76 E) 80

Problem 12. Points B and C lie on $\overline{AD}$. The length of $\overline{AB}$ is 3 times the length of $\overline{BD}$, and the length of $\overline{AC}$ is 7 times the length of $\overline{CD}$. What is the ratio $\frac{AD}{BC}$?

A) 4 B) $\frac{9}{2}$ C) $\frac{13}{2}$ D) 7 E) 8

Problem 13. How many real number solutions does the equation $5^x x^2 + 125 = 5^{x+2} + 5x^2$ have?

A) 0 B) 1 C) 2 D) 3 E) More than 3

Problem 14. Let S be the set of all rational numbers greater than 0 and less than 1 which can be written in simplest form as $\frac{a}{b}$, where a and b are relatively prime positive integers and b is a divisor of 1000. For example, $\frac{1}{2}$, $\frac{3}{250}$, and $\frac{999}{1000}$ are elements of S. What is the sum of all elements of S?

A) 499 B) $\frac{999}{2}$ C) 500 D) $\frac{1001}{2}$ E) 501

Problem 15. Alice rolls a fair six-sided die three times and writes her dice rolls as a three-digit number. For example, if her three rolls are 2, 6, and 5, she writes the number "265." What is the probability that the three-digit number Alice writes is divisible by 7?

A) $\frac{5}{36}$ B) $\frac{31}{216}$ C) $\frac{4}{27}$ D) $\frac{35}{216}$ E) $\frac{1}{6}$

Problem 16. Regular dodecagon $A_1A_2A_3 \dots A_{12}$ has side length 1. What is the area of hexagon $A_1A_3A_5A_7A_9A_{11}$?

A) $4 + 3\sqrt{3}$ B) $\frac{9\sqrt{3}}{2}$ C) $\frac{9 + 6\sqrt{3}}{2}$ D) $6\sqrt{3}$ E) $\frac{18 + 12\sqrt{3}}{2}$

Problem 17. If x and y are real numbers satisfying $x^2 + xy + y^2 = 3$, then which of the following *cannot* be equal to $x^2 + y^2$?

A) 1 B) 2 C) 3 D) 4 E) 5

Problem 18. What is the largest positive integer n such that the number $17^{128} - 1$ is divisible by 2^n?

A) 10 B) 11 C) 12 D) 13 E) None of the preceding

Problem 19. How many of the first 1000 positive integers are there whose base-4 and base-8 representations both contain the digit '0'?

A) 279 B) 300 C) 324 D) 338 E) 352

Problem 20. Triangle ABC has $\angle C = 90^\circ$. Point D is the midpoint of hypotenuse AB, and R_1 and R_2 are the radii of the circumscribed circles of $\triangle ACD$ and $\triangle BCD$, respectively. Given that $BC = a$, $AC = b$, and $AB = c$, what is $\dfrac{R_1}{R_2}$?

A) $\dfrac{a}{b}$ B) $\dfrac{a^2}{b^2}$ C) $\dfrac{b}{a}$ D) $\dfrac{b^2}{a^2}$ E) None of the preceding

Problem 21. Let x and y be positive real numbers such that $\frac{x}{y}+\frac{y}{x}+\frac{x^2}{y^2}+\frac{y^2}{x^2}=28$. What is the value of $\frac{(x+y)^2}{xy}$?

A) 3 B) 7 C) 11 D) 15 E) 19

Problem 22. Positive integers a, b, c, and d satisfy the equation $3^{3a}+3^{4b}+3^{5c}=3^{7d}$. What is the minimum possible value of $a+b+c+d$?

A) 111 B) 194 C) 256 D) 260 E) 278

Problem 23. Let n be the largest positive three-digit integer satisfying the following equations:

$$\text{gcd}(n,20)=10$$
$$\text{gcd}(n+1,21)=7$$

where $\text{gcd}(a,b)$ denotes the greatest common divisor of a and b. What is the sum of the digits of n?

A) 7 B) 9 C) 10 D) 12 E) 15

Problem 24. Let $S = \{-20, -19, -18, \ldots, 19, 20\}$. An integer n is *good* if there exist integers $a, b \in S$ such that $a \neq b$ and $n = ab$. What is the sum of all good integers?

A) -2870 B) -2741 C) -2641 D) -2497 E) 0

Problem 25. Let ABC be a triangle, and let D be the midpoint of $\overline{BC}$. The incircle ω of $\triangle ABC$ intersects $\overline{AD}$ at two distinct points, which trisect the median $\overline{AD}$. Circle ω is tangent to $\overline{BC}$ at point E. Given that $ED = 8$, and D is between points E and C, what is the area of ω?

A) 8π B) $\dfrac{72}{7}\pi$ C) $\dfrac{121}{5}\pi$ D) 49π E) $\dfrac{1008}{17}\pi$

Practice Exam 6

- You have **75 minutes** for **25 problems**.
- There are no penalties for incorrect answers. Answer as many problems as you can; return to the others in the time you have left for the test.

Problem 1. Richard adds up the integers between 1 and 100, inclusive, and Sam adds up the odd integers between 1 and 200, inclusive. What is Sam's sum minus Richard's sum?

A) 99 B) 100 C) 2500 D) 4950 E) 5050

Problem 2. How many prime numbers p are there such that $p + 10$ and $p + 20$ are also prime?

A) 0 B) 1 C) 2 D) 3 E) None of the preceding

Problem 3. Alice has a bag with ten chips, labeled 1, 2, 3, ..., 10. She randomly selects two chips without replacement. What is the probability that the sum of the numbers on the two chips is even?

A) $\frac{1}{4}$ B) $\frac{4}{9}$ C) $\frac{1}{2}$ D) $\frac{5}{9}$ E) $\frac{3}{4}$

Problem 4. Regular octagon $ABCDEFGH$ has side length 2. What is the area of square $ACEG$?

A) 8 B) $8+2\sqrt{2}$ C) $6+4\sqrt{2}$ D) $7+4\sqrt{2}$ E) $8+4\sqrt{2}$

Problem 5. On Monday morning, Jerry drove from his house to work via his usual commute at an average speed of 45 mph. That afternoon, he drove a different route, which covered twice as much distance as his morning route, at an average speed of 60 mph. Given that Jerry spent a total of $1\frac{1}{2}$ hours commuting to and from work, how many minutes did he spend in his morning commute?

A) 30 B) 36 C) 40 D) 45 E) $51\frac{3}{7}$

Problem 6. The five-digit number $\overline{\mathrm{M2M2M}}$ is divisible by the two-digit number $\overline{\mathrm{MM}}$. What is the value of M?

A) 1 B) 2 C) 4 D) 5 E) 8

Problem 7. David rolls a fair six-sided die repeatedly until the sum of the dice rolls he has obtained thus far is 6 or greater. What is the probability that he rolls the die exactly three times?

A) $\frac{23}{108}$ B) $\frac{25}{108}$ C) $\frac{1}{4}$ D) $\frac{5}{18}$ E) None of the preceding

Problem 8. A ball is placed in the lower left corner of a billiards table measuring 8 feet by 4 feet. The ball is struck so that it travels along the dotted path as shown, landing in the center pocket (halfway along the longer side). In feet, which of the following is closest to the total distance the ball travels? Assume the diameter of the ball is negligible.

A) 13.3
B) 13.6
C) 14.0
D) 14.4
E) 14.9

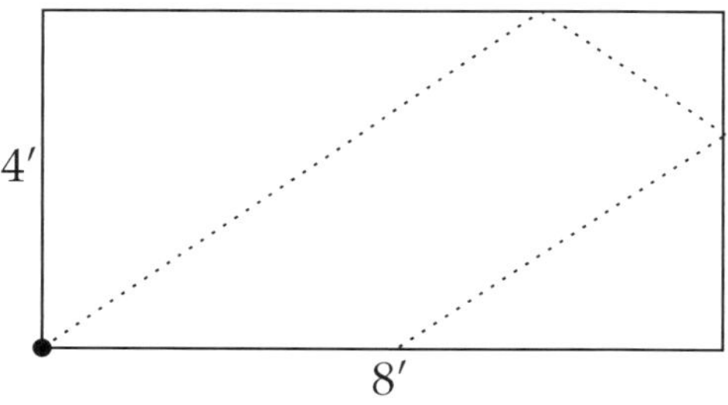

Problem 9. The numbers $\log_2 a$, $\log_2 2020$, and $\log_2 b$ form an arithmetic sequence in that order. What is ab?

A) 2020 B) 4040 C) 2020^2 D) 2×2020^2 E) Cannot be determined

Problem 10. What is the remainder when $10!$ is divided by 99?

A) 9 B) 18 C) 36 D) 54 E) 90

Problem 11. A teacher has 20 identical candies to distribute to four students, Alice, Bob, Carl, and David. However, Bob and David insist that they receive the same number of candies. How many ways can the teacher distribute the candies to his students subject to this constraint, given that each student receives at least one candy?

A) 81 B) 90 C) 100 D) 360 E) 969

Problem 12. Suppose $ABCD$ is a trapezoid such that $\overline{AB} \parallel \overline{CD}$. If $AD = 8$, $DC = 11$, $BC = 15$, and $m\angle C + m\angle D = 270°$, what is the length of $\overline{AB}$?

A) 26 B) 28 C) 30 D) 32 E) 34

Problem 13. If $\sqrt{\underbrace{2019^9 + \cdots + 2019^9}_{m \text{ times}}} = 2019^{19}$, then what is the value of m?

A) 2019^9 B) 2019^{19} C) 2019^{29} D) 2019^{39} E) 2019^{49}

Problem 14. As n ranges over all positive integers, how many possible values can $\gcd(6n + 15, 10n + 21)$ equal? The notation $\gcd(a, b)$ represents the greatest common divisor of a and b.

A) 1 B) 2 C) 4 D) 6 E) Infinitely many

Problem 15. A wall contains four circular dials, each with an arrow facing up. A move consists of selecting two adjacent dials, then rotating them clockwise by 90°. Out of the $4^4 = 256$ possible configurations of the dials, how many configurations can be obtained via a sequence of zero or more moves?

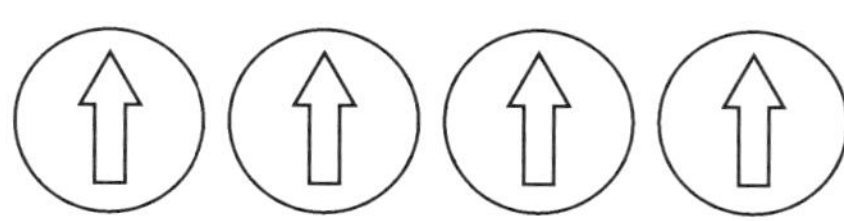

A) 16 B) 32 C) 64 D) 128 E) 256

Problem 16. A trapezoid has vertices at $(0,0)$, $(10,0)$, $(6,5)$, and $(0,5)$. Line ℓ passing through the origin divides the trapezoid into two regions of equal area. What is the slope of line ℓ?

A) $\frac{1}{2}$ B) $\frac{4}{7}$ C) $\frac{10}{17}$ D) $\frac{13}{21}$ E) $\frac{5}{6}$

Problem 17. What is the sum of all real numbers x which satisfy the equation $4^x + 1 = 2^{x+3}$?

A) 0 B) 1 C) 2 D) 4 E) 8

Problem 18. Let $s(n)$ denote the sum of the digits of n, when expressed in base 10. For example, $s(37) = 3 + 7 = 10$. For how many three-digit positive integers n is it true that $s(n) = s(2n)$?

A) 35 B) 36 C) 44 D) 79 E) 80

Problem 19. A bag contains 101 coins, 98 of which are real and three of which are counterfeit. The weights of the counterfeit coins are equal and lighter than that of a real coin. Using a two-pan balance scale, what is the fewest number of weighings needed so that we can find a set of 25 real coins?

A) 2 B) 3 C) 4 D) 5 E) None of the preceding

Problem 20. Triangle ABC is a right triangle with $\angle ABC = 90°$. Point D is on $\overline{BC}$ such that $AD \cdot DC = 2BD \cdot AC$. What is $\dfrac{m\angle DAC}{m\angle BAD}$?

A) 1 B) $\frac{5}{4}$ C) $\frac{4}{3}$ D) $\frac{3}{2}$ E) 2

Problem 21. Suppose $\overline{XX}, \overline{YY}$ and $\overline{ZZ}$ are two-digit whole numbers, where X, Y, and Z are digits. If $X^2+Y^2+Z^2=74$, then how many positive divisors does the number $\overline{XX}^2+\overline{YY}^2+\overline{ZZ}^2$ have?

A) 6 B) 8 C) 12 D) 16 E) Cannot be determined

Problem 22. In $\triangle ABC$, $AB=3$, $BC=4$, and $CA=5$. Let M and N be on $\overline{BC}$ and $\overline{AC}$, respectively, such that line segment $\overline{MN}$ splits $\triangle ABC$ into two regions of equal area, what is the minimum possible value of MN?

A) 1 B) $\sqrt{2}$ C) $\frac{3}{2}$ D) 2 E) $\frac{3\sqrt{2}}{2}$

Problem 23. John writes a sequence of positive integers on a blackboard, beginning with 1. Each term thereafter is the smallest positive integer which cannot be expressed as the product of one or more distinct terms already on the blackboard. The first several terms of John's sequence are 1, 2, 3, 4, 5, 7, 9, 11. How many numbers less than 1,000 appear in John's sequence?
Note: There are 168 prime numbers less than 1,000.

A) 169 B) 180 C) 184 D) 199 E) 203

Problem 24. When expanded out, the polynomial $(x^3 - 20x^2 - 20x + 1)^{10}$ is equal to $a_{30}x^{30} + a_{29}x^{29} + a_{28}x^{28} + \dots + a_1x + a_0$, for integer coefficients $a_{30}, \dots, a_0$. What is $a_{30} + a_{28} + a_{26} + \dots + a_2 + a_0$?

A) $-2^9 \times 19^{10}$ B) $-2^{10} \times 19^9$ C) -38^9 D) $2^9 \times 19^{10}$ E) 38^{10}

Problem 25. A frog starts at "0" on a number line, and wishes to hop to the number "6." Each second, the frog flips a fair coin; if the coin lands heads, the frog hops forward 1 unit; if the coin lands tails, the frog hops forward 2 units (or 1 unit if the frog is already on the number "5"). Which of the following is closest to the expected number of hops the frog takes before it hops to the number "6?"

A) 3.4 B) 4.0 C) 4.2 D) 4.3 E) 4.5

Practice Exam 7

- You have **75 minutes** for **25 problems**.
- There are no penalties for incorrect answers. Answer as many problems as you can; return to the others in the time you have left for the test.

Problem 1. What is the value of the expression $\dfrac{10^{2020}+10^{2018}}{10^{2020}-10^{2018}}$?

A) $\dfrac{1001}{999}$ B) $\dfrac{1010}{1009}$ C) $\dfrac{101}{99}$ D) $\dfrac{11}{9}$ E) 2020

Problem 2. What is the units digit of the number $3^{2020}+7^{2020}$?

A) 0 B) 2 C) 4 D) 6 E) 8

Problem 3. Five people, including a couple and their child, sit in a row containing five seats. In how many ways can they seat themselves, given that the child must sit next to both of his parents?

A) 3 B) 6 C) 12 D) 18 E) 36

Problem 4. The ratio of the corresponding side lengths of two similar triangles is $5:4$, and the perimeter of the greater triangle is 30 cm. What is length the of the shortest side of the smaller triangle, if its side lengths are consecutive even numbers?

A) 2 cm B) 4 cm C) 6 cm D) 8 cm E) 10 cm

Problem 5. If the function f satisfies $f\left(x-\frac{1}{x}\right)=x^2+\frac{1}{x^2}$ for all positive real numbers x, what is the value of $f(2)$?

A) $\frac{1}{2}$ B) $\frac{3}{2}$ C) 3 D) 4 E) 6

Problem 6. A *semiprime* is a positive integer which is the product of two (not necessarily distinct) prime numbers. Let n be the smallest integer greater than 100 such that the numbers n, $n+1$, and $n+2$ are semiprimes. What is the sum of the digits of n?

A) 1 B) 2 C) 3 D) 4 E) 6

Problem 7. Two different whole numbers are randomly selected from the set $\{1, 2, 3, \ldots, 10\}$. What is the probability that their product is an even number?

A) $\frac{2}{9}$ B) $\frac{5}{9}$ C) $\frac{2}{3}$ D) $\frac{3}{4}$ E) $\frac{7}{9}$

Problem 8. A tetrahedron has a base which is an equilateral triangle. The height of the tetrahedron is $2\sqrt{3}$, and the angle between the lateral face and the base of the tetrahedron is $60°$. What is the volume of the tetrahedron?

A) 24 B) 27 C) 36 D) 48 E) 72

Problem 9. Calvin's calculator is broken, and only the '1', '+', and '−' keys work correctly. For example, to key in the number 21, Calvin enters in '$11+11-1$', which requires seven keystrokes. What is the fewest number of keystrokes Calvin needs in order to key in the number 2020?

A) 26 B) 27 C) 28 D) 29 E) 30

Problem 10. For how many positive integers n less than 1000 is the quantity $n^2 + 8n - 85$ a multiple of 101?

A) 0 B) 2 C) 8 D) 9 E) 10

Problem 11. An ant wishes to crawl from point A to point B in the diagram shown below, left. The ant must crawl in the direction of the arrows shown, and may crawl in either direction on the horizontal segments. The ant may not crawl on the same segment more than once. How many different ways can the ant crawl from A to B? One way is shown below, right.

A) 144
B) 216
C) 288
D) 360
E) 384

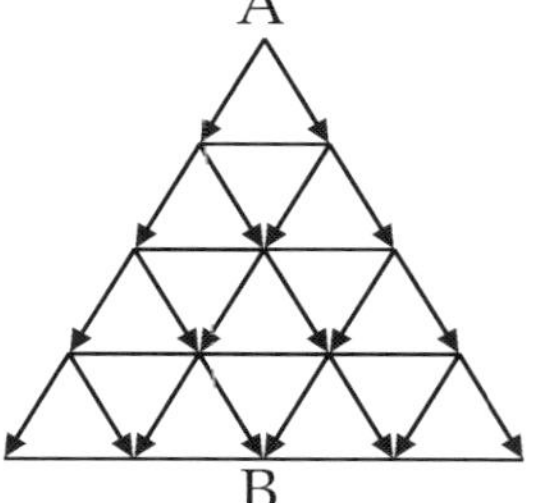

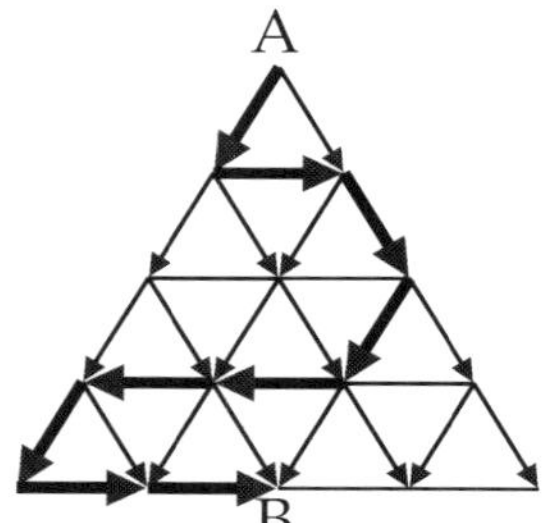

Problem 12. In $\triangle ABC$, point M is the midpoint of side $\overline{BC}$. If $AB = AM = 6$ and $AC = 8$, what is the perimeter of $\triangle ABC$?

A) $14 + \sqrt{14}$ B) $14 + 2\sqrt{14}$ C) 22 D) 24 E) $14 + \frac{8\sqrt{21}}{3}$

Problem 13. What is the area of the region in the xy-plane bounded by the graphs of $y = |x - 4| + |x - 6|$ and $y = 10$?

A) 24 B) 28 C) 32 D) 36 E) 48

Problem 14. The sum of three positive integers is 314. At most how many zeros does the product of these three integers end with, when written in base 10?

A) 5 B) 6 C) 7 D) 8 E) None of the preceding

Problem 15. A positive divisor of $20!$ is randomly chosen. What is the probability that it is divisible by 100?

A) $\frac{8}{15}$ B) $\frac{51}{95}$ C) $\frac{54}{95}$ D) $\frac{14}{19}$ E) $\frac{73}{95}$

Problem 16. Let P be a point inside equilateral triangle ABC such that $m\angle APB = 150^\circ$, $AP = 2\sqrt{3}$ and $BP = 2$. What is PC?

A) 3 B) $2\sqrt{3}$ C) $\sqrt{14}$ D) 4 E) $3\sqrt{3}$

Problem 17. If n is a positive integer such that $\left(1-\frac{1}{2^2}\right)\cdot\left(1-\frac{1}{3^2}\right)\cdot\left(1-\frac{1}{4^2}\right)\cdots\left(1-\frac{1}{n^2}\right) = \frac{n+1}{16}$, then what is the value of n?

A) 4 B) 8 C) 9 D) 12 E) 16

Problem 18. Given a positive integer n, let $s(n)$ denote the sum of the digits of n. For example, $s(123) = 6$. For how many of the first 1000 positive integers n is it true that there exists a positive integer $k \geq 2$ which is not a power of 10 such that $s(n) = s(kn)$?

A) 983 B) 994 C) 996 D) 997 E) 1000

Problem 19. Two real numbers a and b are randomly and uniformly chosen from the interval $[0, 2]$. What is the probability that a, b, and 1 are the side lengths of a triangle?

A) $\frac{1}{4}$ B) $\frac{3}{8}$ C) $\frac{1}{2}$ D) $\frac{5}{8}$ E) $\frac{3}{4}$

Problem 20. A cube has side length 4 inches. A solid is formed by slicing eight congruent tetrahedra from the cube as shown. What is the surface area of the solid, in square inches?

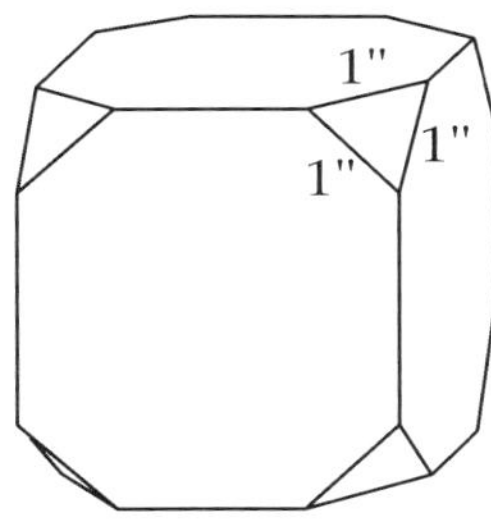

A) $84 + 2\sqrt{3}$ B) $90 + 2\sqrt{3}$ C) $90 + 4\sqrt{3}$ D) $102 - 2\sqrt{3}$ E) None of the preceding

Problem 21. Suppose x and y are real numbers such that $\sqrt{x^2+1}+\sqrt{y^2+4}=7$. What is the maximum possible value of $x+y$?

A) 6   B) $2\sqrt{10}$ C) $\sqrt{42}$ D) 7   E) None of the preceding

Problem 22. The average value of the pennies, nickels, dimes, and quarters in Qi's pocket is exactly 10 cents. Given that she has at least one of each type of coin, what is the minimum possible number of coins in Qi's pocket?

A) 8   B) 9   C) 11   D) 12   E) 13

Problem 23. For some positive integer x, define P_i to be the point $(x+i, 2^{x+i})$. If the area of convex hexagon $P_0P_1P_2P_3P_4P_5$ is 144, what is the value of x?

A) 1   B) 2   C) 3   D) 4   E) 5

Problem 24. Let R denote the set of all complex numbers z with the property that there exist real numbers $a, b \in [0, 1]$ such that $z^2 + az + b = 0$. What is the area of R, when graphed in the complex plane?

A) $\frac{\sqrt{3}}{4} + \frac{\pi}{6}$ B) $\frac{\sqrt{3}}{2} + \frac{\pi}{6}$ C) $\frac{\pi}{2}$ D) $\sqrt{3} + \frac{2\pi}{3}$ E) 2π

Problem 25. A bag contains five marbles of different colors: red, green, blue, yellow, and white. José randomly selects one marble from the bag, discards it, then adds a red marble into the bag. He repeats this process of selecting one marble, discarding it, then adding a red marble into the bag a total of five times. The probability that the fifth marble José selects from the bag is red can be written in the form $\frac{m}{n}$, where m and n are relatively prime positive integers. What is the remainder when $m + n$ is divided by 1000?

A) 126 B) 226 C) 429 D) 501 E) 629

Practice Exam 8

- You have **75 minutes** for **25 problems**.
- There are no penalties for incorrect answers. Answer as many problems as you can; return to the others in the time you have left for the test.

Problem 1. Patrick the master gardener has 100 lbs of tomatoes that are 90% water by weight. He dried his tomatoes in the sun until they were 80% water by weight. How many pounds do Patrick's sun-dried tomatoes weigh now?

A) 50 B) 60 C) 80 D) $88.\overline{8}$ E) 90

Problem 2. How many integers n satisfy the double inequality $\frac{1}{4} \leq \frac{n}{2018} \leq \frac{1}{3}$?

A) 160 B) 164 C) 167 D) 168 E) 169

Problem 3. Rachel has two \$1 bills, two \$5 bills, two \$10 bills, and two \$20 bills. Using any combination of one or more of these bills, how many different monetary amounts can Rachel form? For example, she can form \$27 using one \$20 bill, one \$5 bill, and two \$1 bills.

A) 40 B) 41 C) 42 D) 44 E) 45

Problem 4. A rectangular fish tank has base dimensions 4 feet by 2 feet, and height 3 feet. The tank is initially half full. A solid steel ball of diameter 1 foot is dropped into the tank, which sinks to the bottom. By approximately how many inches does the water level in the tank rise? **Note:** 1 foot equals 12 inches.

A) 0.3 B) 0.5 C) 0.6 D) 0.8 E) 1.2

Problem 5. Let n be a positive integer. How many elements are in the set $\left\{a \in \mathbb{N} : |\sqrt{a} - n| < \frac{1}{2}\right\}$?

A) $n-1$ B) $n+1$ C) $2n-1$ D) $2n$ E) $n(n-1)$

Problem 6. Let $N = 20! + 21! + 22! + 23!$. Let p be the largest prime divisor of N. What is the sum of the digits of p?

A) 2 B) 5 C) 7 D) 10 E) 11

Problem 7. Real numbers a, b, and c are randomly and uniformly chosen from the interval $[-1, 1]$. What is the probability that the area of the triangle formed by the points $(a, |a|)$, $(b, |b|)$, and $(c, |c|)$ is positive?

A) $\frac{1}{2}$ B) $\frac{5}{8}$ C) $\frac{3}{4}$ D) $\frac{7}{8}$ E) 1

Problem 8. Square $ABCD$ has side length 1. It is rotated 45° about A to form a new square, $AB'C'D'$. What is the area of the region common to both squares?

A) $\frac{4\sqrt{2}-5}{2}$ B) $\frac{\sqrt{2}}{4}$ C) $\frac{3}{8}$ D) $\sqrt{2}-1$ E) $\frac{1}{2}$

Problem 9. How many integer solutions x are there to the equation $(x^2 - 3x + 1)^{x+1} = 1$?

A) 0 B) 1 C) 2 D) 3 E) 4

Problem 10. What is the remainder when $4^{2018} + 6^{2018}$ is divided by 25?

A) 2 B) 4 C) 18 D) 12 E) 24

Problem 11. A bag contains nine chips labeled 1, 2, 3, ..., 9. Sean randomly selects two chips from the bag, without replacement. What is the expected value of the product of the numbers on the two chips Sean selects?

A) 24 B) $24\frac{1}{6}$ C) $24\frac{5}{6}$ D) 25 E) $25\frac{1}{2}$

Problem 12. In triangle ABC, $m\angle ABC = 2m\angle BCA$, $BC = 7$, and $AB = 9$. What is AC?

A) 8 B) 10 C) $\frac{21}{2}$ D) 12 E) $\frac{27}{2}$

Problem 13. What is the sum of all real solutions x to the equation $\sqrt{4x - 5} + \sqrt[3]{6 - 4x} = 1$?

A) $\frac{5}{4}$ B) $\frac{7}{2}$ C) $\frac{9}{2}$ D) $\frac{25}{4}$ E) $\frac{34}{5}$

Problem 14. Alice counts the first sixty positive multiples of 2: 2, 4, 6, 8, …, 120. Similarly, Bob counts the first sixty positive multiples of 3, and Charlie counts the first sixty positive multiples of 5. How many numbers were counted by at least one person?

A) 88 B) 120 C) 136 D) 140 E) 180

Problem 15. Let $s(n)$ denote the sum of the digits of n, when written in base 10. For example, $s(123) = 6$ and $s(2020) = 4$. How many positive integers n less than 10^4 have the property that $s(n) > 10$?

A) 8998 B) 8999 C) 9002 D) 9003 E) 9004

Problem 16. Triangle ABC is isosceles with $AB = AC = 5$ and $BC = 8$. Point P is inside $\triangle ABC$ such that the product of the areas of $\triangle APB$, $\triangle APC$, and $\triangle BPC$ is maximized. What is PB?

A) 3 B) $\sqrt{10}$ C) $2\sqrt{3}$ D) $\sqrt{17}$ E) $2\sqrt{5}$

Problem 17. Suppose x and y are real numbers. What is the minimum possible value of $x^2 + y^2 + xy - 4x - 5y$?

A) -7 B) -6 C) -4 D) -3 E) 0

Problem 18. For how many ordered pairs (p, q) of prime numbers is the expression $p^2 + 3pq + q^2$ a perfect square?

A) 0 B) 2 C) 4 D) 6 E) 8

Problem 19. Ben rolls a fair six-sided die repeatedly until the arithmetic mean (average) of all his dice rolls is 3.5 or greater, at which point he stops. What is the probability that Ben rolls the die at most three times?

A) $\frac{2}{3}$ B) $\frac{37}{54}$ C) $\frac{17}{24}$ D) $\frac{3}{4}$ E) $\frac{55}{72}$

Problem 20. Circles ω_1 and ω_2, each with radius 1, are drawn in the plane such that the distance between the centers of ω_1 and ω_2 is 1. Circles ω_1 and ω_2 intersect at two distinct points A and B. Points P and Q are on ω_1 and ω_2 respectively, such that P, A, and Q are collinear and A is between P and Q. If $QA = 2AP$, what is the length of segment PQ?

A) $\frac{9}{7}$ B) $\frac{3\sqrt{21}}{7}$ C) $\frac{3\sqrt{2}}{2}$ D) $\frac{3\sqrt{33}}{8}$ E) $\frac{3\sqrt{26}}{7}$

Problem 21. A function $f : \mathbb{Z} \to \mathbb{Z}$ satisfies $f(x+y) = 4xy + f(x-y)$ for all integers x and y. Given that $f(0) = 0$ and $f(1) = 3$, what is the value of $f(19) + f(20)$?

A) 760 B) 761 C) 762 D) 763 E) Cannot be determined

Problem 22. Let $N = 3 \times 7 \times 13 \times 17 \times ... \times 993 \times 997$ denote the product of all positive integers less than 1000 whose units digit is either 3 or 7. What is the remainder when N is divided by 1000?

A) 1 B) 251 C) 501 D) 751 E) 991

Problem 23. The polynomial $P(z) = z^{12} + z^8 + z^4 + 1$ has 12 distinct complex roots. Let R be the convex dodecagon whose vertices are the roots of $P(z)$ when plotted in the complex plane. Which of the following is closest to the area of R?

A) 2.9 B) 3.1 C) 3.2 D) 3.4 E) 3.6

Problem 24. Alice has an unusual coin which has two sides, heads and tails. The probability that the first flip lands heads is 1, and the probability that the second flip lands tails is 1. For all $n \geq 3$, the probability that the nth flip lands heads is $\frac{H_{n-1}}{n-1}$, where H_k denotes the number of heads flipped out of the first k coin flips. If Alice flips this coin a total of 2020 times, what is the probability that exactly 1010 flips land heads?

A) $\frac{1}{2021}$ B) $\frac{1}{2020}$ C) $\frac{1}{2019}$ D) $\frac{1}{2018}$ E) None of the preceding

Problem 25. Equiangular hexagon $ABCDEF$ has $AB = CD = EF = 28$ and $BC = DE = FA = 14$. Points M, N, O, P, Q, and R are the midpoints of sides $\overline{AB}$, $\overline{BC}$, $\overline{CD}$, $\overline{DE}$, $\overline{EF}$, and $\overline{FA}$, respectively. There exists a unique circle ω which is tangent to line segments $\overline{MN}$, $\overline{OP}$, and $\overline{QR}$. What is the area of ω?

A) 189π B) $\frac{833\pi}{4}$ C) $\frac{700\pi}{3}$ D) $\frac{1029\pi}{4}$ E) 336π

Practice Exam 9

- You have **75 minutes** for **25 problems**.
- There are no penalties for incorrect answers. Answer as many problems as you can; return to the others in the time you have left for the test.

Problem 1. If $\dfrac{2}{1300x-2600} = \dfrac{1}{650} - \dfrac{1}{651}$, then what is the value of x?

A) 650 B) 651 C) 653 D) 1302 E) 1303

Problem 2. Given a positive integer n, let $s(n)$ denote the sum of the digits of n. For example, $s(37) = 3 + 7 = 10$. Given that $s(n) = 99$, what is the sum of all possible values of $s(n + 1)$?

A) 100 B) 191 C) 192 D) 606 E) 909

Problem 3. Marek rolls a fair six-sided die with faces labeled 1, …, 6, and Nisha rolls a fair eight sided die, with faces labeled 1, …, 8. What is the probability that Nisha's roll is greater than Marek's roll?

A) $\frac{9}{16}$ B) $\frac{31}{48}$ C) $\frac{2}{3}$ D) $\frac{11}{16}$ E) $\frac{3}{4}$

Problem 4. Triangle ABC is an isosceles triangle with $\mathrm{AB} = \mathrm{BC}$. Point D is inside $\triangle\mathrm{ABC}$ such that $\mathrm{AD} = \mathrm{DC}$. If $\mathrm{m}\angle\mathrm{ABC} = 50^\circ$ and $\mathrm{m}\angle\mathrm{ADC} = 130^\circ$, then $\mathrm{m}\angle\mathrm{BAD} = ?$

A) 35° B) 40° C) 45° D) 50° E) 55°

Problem 5. Seven students each took a test worth 100 points. The mean, median, and unique mode of the students' scores were all equal to 90 points. Given that each student scored a whole number of points between 0 and 100 inclusive, what is the lowest possible score that a student could have received?

A) 60 B) 61 C) 63 D) 64 E) 65

Problem 6. Suppose n is a positive integer such that $2^{5n+1} + 5^{n+k}$ is divisible by 27. Which of the following could be the value of k?

A) 1 B) 2 C) 4 D) 5 E) 7

Problem 7. Two ants start at opposite vertices of a unit cube. Each minute, each ant moves to a randomly chosen vertex of the cube adjacent from its current vertex. What is the probability that, after 3 minutes, both ants are on the same vertex of the cube?

A) 0 B) $\frac{1}{8}$ C) $\frac{1}{6}$ D) $\frac{1}{4}$ E) $\frac{1}{3}$

Problem 8. Let $ABCD$ be an isosceles trapezoid with $AB = DC$ and bases $AD = 12$ and $BC = 30$. Lines $\overleftrightarrow{AB}$ and $\overleftrightarrow{CD}$ intersect at point P outside of the trapezoid with $AP = 8$. What is the perimeter of trapezoid $ABCD$?

A) 60 B) 66 C) 72 D) 84 E) 92

Problem 9. The largest real solution x to the equation $\log_2 x + \log_4 x + \log_8 x = (\log_2 x)(\log_4 x)(\log_8 x)$ lies in which of the following intervals?

A) $[1,2)$ B) $[2,4)$ C) $[4,8)$ D) $[8,16)$ E) $[16,32)$

Problem 10. The sequence (a_n) is defined by $a_1 = a_2 = 1$, and $a_n = 2a_{n-1} + 2a_{n-2}$ for $n \geq 3$. What is the remainder when $a_1 + a_2 + a_3 + ... + a_{2020}$ is divided by 16?

A) 0 B) 2 C) 4 D) 8 E) 10

Problem 11. What is the coefficient of x^6 in the expansion of $(1 + x + x^2 + x^3)^4$?

A) 36 B) 38 C) 40 D) 44 E) 48

Problem 12. A circle with center O is tangent to sides $\overline{AB}$, $\overline{BC}$, and $\overline{AC}$ of triangle ABC at points K, L, and M, respectively. If $AM = 8$, $AB = 11$, and $BC = 10$, what is the area of triangle ABC?

A) $12\sqrt{3}$ B) $9\sqrt{6}$ C) 24 D) $12\sqrt{21}$ E) None of the preceding

Problem 13. The function f is piecewise defined on the positive integers as follows:

$$f(x) = \begin{cases} x+1 & x \text{ is odd} \\ \dfrac{x}{2} & x \text{ is even} \end{cases}$$

What is the value of $f(f(1)) + f(f(2)) + f(f(3)) + \ldots + f(f(2020))$?

A) 825650 B) 843350 C) 867050 D) 893850 E) 944350

Problem 14. Let n be the smallest positive integer with the property that the least common multiple of $20!$ and n is $21!$. How many positive divisors does n have?

A) 4 B) 24 C) 27 D) 30 E) 40

Problem 15. An 8-question true-false test is scored such that for every integer $1 \leq n \leq 8$, question n is worth n points, so that the maximum score is 36 points. A student randomly guesses either true or false on every question. What is the probability that the student scores at least 50% of all possible points on the test?

A) $\frac{131}{256}$ B) $\frac{33}{64}$ C) $\frac{133}{256}$ D) $\frac{67}{128}$ E) $\frac{135}{256}$

Problem 16. Unit circles ω_1 and ω_2 are drawn in the plane such that the distance between their centers is $\frac{3}{2}$. Points A and B are on ω_1 and ω_2, respectively, such that $\overline{AB}$ is trisected by the intersection points with ω_1 and ω_2 as shown below. What is the length of $\overline{AB}$?

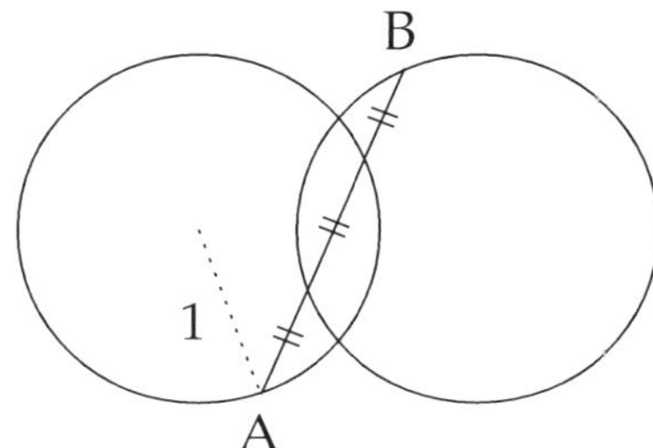

A) $\frac{3}{2}$ B) $\frac{\sqrt{15}}{2}$ C) $\frac{3\sqrt{2}}{2}$ D) $\frac{\sqrt{21}}{2}$ E) $\frac{7}{3}$

Problem 17. A triangle has vertices $(0,0)$, $(2,4)$, and (x,x^2) where $0 < x < 2$. What is the maximum possible area of this triangle?

A) $\frac{1}{2}$ B) $\frac{3}{4}$ C) 1 D) 2 E) None of the preceding

Problem 18. For some positive integer k, the last two digits of $11^k + 13^k$ when written in base 10 are "24." Which of the following could be the value of k?

A) 80 B) 81 C) 82 D) 83 E) 84

Problem 19. Five children initially have 5, 10, 15, 20, and 25 Halloween candies. On every move, one child (who has at least four candies) gives one candy to each of the other four children. What is the minimum number of moves needed so that every child has an equal number of candies?

A) 8 B) 10 C) 12 D) 15 E) It is impossible for the children to have an equal number of candies.

Problem 20. Triangle ABC has $m\angle C = 90^\circ$. Point H is the foot of the perpendicular from C to $\overline{AB}$, and point D is on $\overline{AB}$ such that $\overline{CD}$ bisects $\angle HCA$. Given that $HD = 1$ and $BC = 5$, what is $AD + AC$?

A) 3 B) 4 C) 5 D) 6 E) None of the preceding

Problem 21. David is playing a game with three fair six-sided dice. He rolls each die once. If the sum of the numbers on the top faces is 11 or greater, David wins. Otherwise, he rerolls one die with the lowest number on top. If the sum of the numbers on the top faces after rerolling is 11 or greater, David wins; otherwise, he loses. The probability that David wins can be written as $\frac{m}{n}$ for relatively prime positive integers m and n. What is $m + n$?

A) 593 B) 641 C) 649 D) 793 E) 797

Problem 22. A sequence (a_n) is defined recursively with $a_1 = a_2 = 1$, and $a_n = 2a_{n-1} + a_{n-2}$ for all integers $n \geq 3$. What is the value of

$$\sum_{i=1}^{\infty} \frac{a_i}{3^i}?$$

A) 1 B) $\frac{3}{2}$ C) 2 D) $\frac{5}{2}$ E) The series diverges.

Problem 23. Let R be the region in the xy-plane bounded by the parabolas $y = x^2 - 4$ and $y = -x^2 + 4$. A circle centered at the origin is drawn, completely contained in R. What is the maximum possible area of this circle?

A) $\frac{13}{4}\pi$ B) $\frac{7}{2}\pi$ C) $\frac{15}{4}\pi$ D) 4π E) 8π

Problem 24. Triangle ABC has $AB = 13$, $BC = 14$, and $AC = 15$. Let J_A be the center of the A-excircle of triangle ABC. What is AJ_A?
Note: The A-*excircle* of $\triangle ABC$ is the circle in the exterior of $\triangle ABC$ which is tangent to $\overline{BC}$ and the extensions of $\overline{AB}$ and $\overline{AC}$.

A) $6\sqrt{15}$ B) $\sqrt{541}$ C) $9\sqrt{7}$ D) 24 E) $3\sqrt{65}$

Problem 25. How many ordered 11-tuples $(a_0, a_1, a_2, \ldots, a_{10})$ of integers satisfy the equation

$$a_0 + 2a_1 + 2^2a_2 + \ldots + 2^{10}a_{10} = 2020$$

where $0 \leq a_i \leq 2$ for all $0 \leq i \leq 10$?

A) 27 B) 33 C) 39 D) 45 E) 51

Practice Exam 10

- You have **75 minutes** for **25 problems**.
- There are no penalties for incorrect answers. Answer as many problems as you can; return to the others in the time you have left for the test.

Problem 1. What is the value of the following expression?

$$\frac{3^{20} - 3^{10}}{(3^5 - 1)(3^5 + 1)}$$

A) 1 B) 3^{10} C) 3^{15} D) 3^{20} E) None of the preceding

Problem 2. Maria was given a positive integer, and was told to multiply it by 7, add 8 to the result, then multiply the resulting number by 9. However, she forgot which order to perform these operations in, so she randomly performed these three operations and obtained an incorrect answer of 245. What answer should Maria have obtained?

A) 135 B) 198 C) 261 D) 324 E) 387

Problem 3. How many four-digit positive integers are there such that all four digits are prime, and the sum of the digits is even?

A) 81 B) 136 C) 144 D) 150 E) 168

Problem 4. In the figure, points E, A, and B are collinear, $\mathrm{AB} = \mathrm{AC}$, and $\mathrm{AD} = \mathrm{DC}$. Given that $\mathrm{m}\angle\mathrm{EAC} + \mathrm{m}\angle\mathrm{ADC} = 230^\circ$, what is $\mathrm{m}\angle\mathrm{DCB}$?

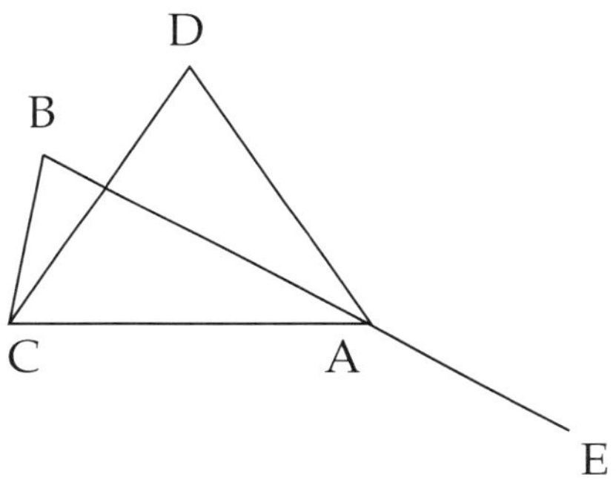

A) 10° B) 15° C) 20° D) 25° E) 30°

Problem 5. Ten children sit around a circle. They count the numbers 1, 2, 3, ...in clockwise order, starting with the oldest child, who says the number '1.' Once a child says a number which is a multiple of 5 or 7, that child leaves the circle and the next child in the circle continues with the next number. They count until only one child remains, at which point they stop counting. What is the last number counted?

A) 25 B) 28 C) 30 D) 35 E) 40

Problem 6. Let a, b and c be distinct prime numbers such that $a(c-b) = 18$ and $b(c-a) = 40$. What is $a+b+c$?

A) 13 B) 17 C) 19 D) 21 E) None of the preceding

Problem 7. Hannah has a bag containing two A tiles, two H tiles, and two N tiles. She selects tiles one at a time from the bag, without replacement. What is the probability that the first three tiles Hannah selects are A, H, and N, in some order?

A) $\frac{1}{6}$ B) $\frac{1}{5}$ C) $\frac{1}{4}$ D) $\frac{1}{3}$ E) $\frac{2}{5}$

Problem 8. The right hexagonal prism shown below has all edges of length 1. An ant crawls from point A to point B along the surface of the hexagonal prism. What is the shortest possible distance the ant can crawl?

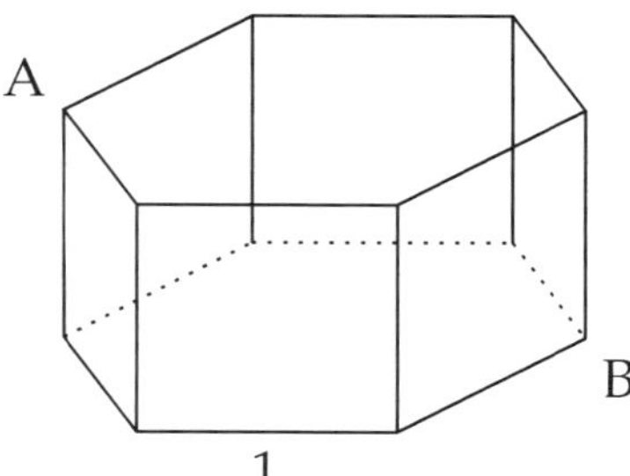

A) $\sqrt{5+\sqrt{3}}$

B) $1+\sqrt{3}$

C) $\sqrt{5+2\sqrt{3}}$

D) 3

E) $\sqrt{2}+\sqrt{3}$

Problem 9. The cubic polynomial $x^3 + 10x^2 - 32x - 384$ has exactly two distinct integer roots r and s. What is $r + s$?

A) -10 B) -4 C) -2 D) 2 E) 10

Problem 10. The number 2^{100} is a 31-digit number whose leftmost digit is 1. How many of the numbers 2^1, 2^2, 2^3, ..., 2^{99} have the property that its leftmost digit is 1?

A) 29 B) 30 C) 31 D) 32 E) None of the preceding

Problem 11. Let $S = \{1,2,3,4,5\}$. How many ordered triples of sets (A,B,C) are there such that $A \cup B \cup C = S$ and $|A \cap B| = |B \cap C| = |C \cap A| = 1$?

A) 405 B) 465 C) 540 D) 585 E) 945

Problem 12. $ABCD$ is a trapezoid with $\overline{AD} \parallel \overline{BC}$, $\overline{AB} \perp \overline{AD}$, $AB = 8$, and $AD = 6$. Given that point E is on $\overline{BC}$ such that $DC = EC = 10$, what is the area of $ABED$?

A) 32 B) 36 C) 40 D) 48 E) 72

Problem 13. The real number $\sqrt{19 - 8\sqrt{3}}$ can be expressed in the form $a + b\sqrt{3}$ where a and b are integers and a is positive. What is $a + b$?

A) 3 B) 4 C) 6 D) 7 E) 8

Problem 14. What is the remainder when $2^{2019} + 3^{2019} + 4^{2019} + 5^{2019} + 6^{2019}$ is divided by 13?

A) 1 B) 3 C) 6 D) 9 E) 11

Problem 15. How many ordered triples (x, y, z) of positive integers satisfy the equation $xyz = 2020$?

A) 12 B) 18 C) 24 D) 48 E) 54

Problem 16. In triangle ABC, let F be a point on $\overline{AB}$, let D be a point on $\overline{BC}$ such that $2BF = AF = 10$ and $BD = 2DC$. Point E is on $\overline{AC}$ such that $EF \perp ED$, $EC = 2AE$, and $ED = 12$. What is the area of $\triangle ABC$?

A) 120 B) 144 C) 150 D) 162 E) 216

Problem 17. Suppose that a, b, and c are positive real numbers such that $ab = a + b + 20$, $bc = b + c + 17$, and $ac = a + c + 41$. What is $a + b + c$?

A) 15 B) 16 C) 19 D) 22 E) None of the preceding

Problem 18. Carl was reading a 500-page book but forgot what page he was on. Somehow, he remembered the page number he was on is a palindrome (a number which reads the same forwards and backwards), the previous page number is a multiple of 9, and the page number before that is a multiple of 11. Let N be the page Carl was on. What is the tens digit of N?

A) 0 B) 1 C) 3 D) 4 E) 6

Problem 19. Binaryland uses a currency consisting of only integer coin denominations which are a power of 2 (1 cent, 2 cents, 4 cents, and so on). How many different monetary amounts less than or equal to 1000 cents can be made using exactly five coins? For example, 547 cents can be made using one 512-cent coin, one 32-cent coin, and three 1-cent coins.

A) 610 B) 614 C) 625 D) 633 E) 637

Problem 20. $ABCD$ is a convex quadrilateral with $m\angle ABC = m\angle ADC = 90°$. Given that $m\angle BAC = 40°$, $m\angle CAD = 20°$, and $BD = 6$, what is AC?

A) $4\sqrt{3}$ B) 6 C) $6\sqrt{3}$ D) 8 E) $8\sqrt{3}$

Problem 21. What is the maximum possible value of $x + y$ for positive integers x and y that satisfy the following equation?

$$\log_2(\log_{2^x}(\log_{2^y} 2^{400})) = 0$$

A) 53 B) 102 C) 201 D) 400 E) None of the preceding

Problem 22. In how many ways can a blank 3×3 grid be filled with the integers from 1 to 9 so that squares containing consecutive integers are adjacent (i.e., share a common edge)? For example, grids **A** and **B** satisfy the given conditions, while **C** does not because the squares containing 1 and 2 are not adjacent.

A

9	8	7
2	1	6
3	4	5

B

7	8	9
6	5	4
1	2	3

C

1	3	2
6	5	4
7	8	9

A) 24 B) 32 C) 36 D) 40 E) 48

Problem 23. Regular hexagon $P_1P_2P_3P_4P_5P_6$ is drawn in the plane. All lines $\overleftrightarrow{P_iP_j}$, where $1 \leq i < j \leq 6$, are drawn in the same plane. How many triangles of positive, but finite, area can be drawn whose sides are contained within these lines? One example triangle is shown below.

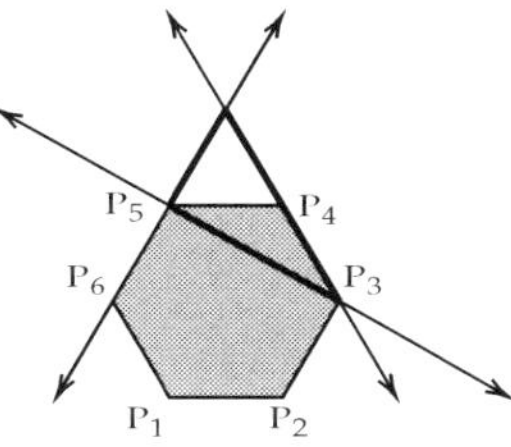

A) 244 B) 245 C) 304 D) 305 E) 455

Problem 24. In triangle ABC, point D is the midpoint of $\overline{AB}$ and point E is on $\overline{AC}$ such that $AC = 3AE$. Circle ω passes through points B, C, E, and D. Given that $m\angle EBC = 90^\circ$, what is $\frac{BE^2}{BC^2}$?

A) $\frac{3}{2}$ B) $\frac{4}{3}$ C) $\frac{5}{3}$ D) $\frac{9}{4}$ E) 2

Problem 25. Zach rolls a fair six-sided die repeatedly until he rolls the same number three times consecutively (e.g., 333 or 666), at which point he stops. What is the expected number of times Zach rolls the die?

A) 36 B) 37 C) 42 D) 43 E) 45

MT PRESS

Answers

1. A
2. D
3. E
4. E
5. E
6. D
7. D
8. D
9. C
10. C
11. D
12. A
13. D
14. B
15. A
16. B
17. E
18. C
19. A
20. B
21. C
22. C
23. C
24. E
25. D

Answers for Practice Exam 2.

1. D
2. B
3. B
4. D
5. C
6. D
7. C
8. B
9. B
10. C
11. D
12. D
13. C
14. B
15. D
16. B
17. E
18. C
19. D
20. E
21. D
22. D
23. E
24. A
25. B

1. A
2. A
3. B
4. A
5. A
6. D
7. C
8. C
9. C
10. C
11. D
12. B
13. B
14. A
15. D
16. D
17. B
18. C
19. A
20. A
21. B
22. D
23. B
24. B
25. B

Answers for Practice Exam 4.

1. B
2. E
3. D
4. B
5. E
6. D
7. E
8. C
9. D
10. C
11. D
12. B
13. C
14. E
15. D
16. D
17. C
18. C
19. E
20. B
21. B
22. B
23. B
24. E
25. D

1. B
2. D
3. D
4. B
5. D
6. E
7. B
8. D
9. E
10. D
11. D
12. E
13. D
14. B
15. A
16. C
17. A
18. B
19. A
20. C
21. B
22. E
23. D
24. D
25. B

Answers for Practice Exam 6.

1. D
2. B
3. B
4. E
5. B
6. D
7. B
8. D
9. C
10. D
11. A
12. B
13. B
14. B
15. C
16. C
17. A
18. E
19. A
20. E
21. C
22. D
23. C
24. D
25. C

MT MATHTOPIA PRESS

1. C
2. B
3. C
4. C
5. E
6. D
7. E
8. A
9. B
10. d
11. E
12. B
13. E
14. B
15. B
16. D
17. B
18. C
19. D
20. B
21. B
22. E
23. B
24. A
25. B

Answers for Practice Exam 8.

1. A
2. D
3. D
4. D
5. D
6. A
7. C
8. D
9. E
10. A
11. B
12. D
13. D
14. D
15. D
16. D
17. A
18. B
19. C
20. B
21. D
22. A
23. A
24. C
25. C

1. C
2. D
3. A
4. B
5. B
6. B
7. A
8. B
9. D
10. A
11. D
12. D
13. D
14. E
15. E
16. D
17. C
18. B
19. B
20. C
21. E
22. A
23. C
24. E
25. E

Answers for Practice Exam 10.

1. B
2. C
3. B
4. D
5. B
6. C
7. E
8. C
9. C
10. A
11. E
12. A
13. A
14. E
15. E
16. D
17. C
18. D
19. D
20. A
21. C
22. D
23. A
24. C
25. D

Solutions

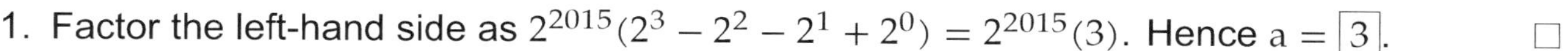

1. Factor the left-hand side as $2^{2015}(2^3 - 2^2 - 2^1 + 2^0) = 2^{2015}(3)$. Hence $a = \boxed{3}$. □

2. We can write the number as $(5^3)^4 \times (2^6)^2 = 5^{12} \times 2^{12} = 10^{12}$, which has $\boxed{13}$ digits. □

3. There are five choices for the entrée. There are five choices for the appetizer (either order one of four appetizers, or none at all), and four choices for the dessert (either order one of three desserts, or none at all). By the multiplication rule, the number of meal combinations is $5 \times 5 \times 4 = \boxed{100}$. □

4. All $\boxed{5}$ shapes are possible:

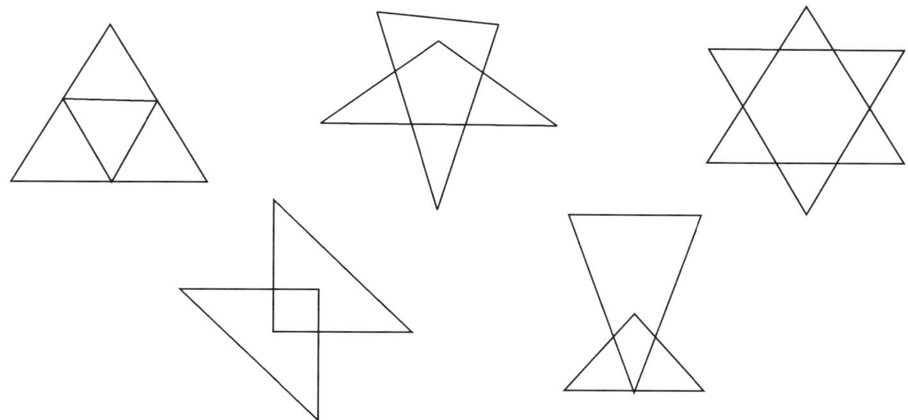

□

5. Let $a \le b \le c \le d \le e$ be the heights of the five students. We are given $a+b+c = 3\times 58 = 174$, $c+d+e = 3 \times 70 = 210$, and $a+b+c+d+e = 5 \times 63 = 315$. Adding the first two equations and subtracting the third, we obtain $c = 174 + 210 - 315 = 69$, so the median height is $\boxed{69}$ inches. □

6. We have $|n^2 - 6n + 5| = |(n-5)(n-1)|$, so in order for this expression to be prime, either $n-5$ or $n-1$ should be ± 1, and the other should be $\pm p$ for some prime p. Testing, we see that $n = 0$, $n = 2$, $n = 4$, and $n = 6$ all work, so the answer is $\boxed{4}$. □

7. We want the probability of obtaining 3 or more heads. The probability is

$$\frac{\binom{6}{3} + \binom{6}{4} + \binom{6}{5} + \binom{6}{6}}{2^6} = \frac{20 + 15 + 6 + 1}{64} = \boxed{\frac{21}{32}}$$

. □

8. Suppose the edge lengths are a, $\frac{3}{2}a$, and $2a$, where a is an integer. Since $\frac{3}{2}a$ is an integer, a must be even, so let $a = 2k$ for some positive integer k. Then the dimensions are $2k$, $3k$, and $4k$, and the volume is $24k^3$. The only answer choice of the form $24k^3$ for some integer k is $\boxed{192}$. □

9. Let the terms of the first sequence be 1, $1+x$, $1+2x$, ..., $1+19x$, and let the terms of the second sequence be 10, $10+y$, $10+2y$, ..., $10+9y$. The sum of the terms of these two sequences are equal, so we have

$$1+(1+x)+(1+2x)+\ldots+(1+19x) = 10+(10+y)+(10+2y)+\ldots+(10+9y)$$
$$20+\frac{19\cdot 20}{2}x = 100+\frac{9\cdot 10}{2}y$$
$$38x-9y = 16$$

Since x and y are positive integers, we want to solve the Diophantine equation $38x-9y=16$ while minimizing $x+y$. Taking this equation modulo 9, we obtain $2x \equiv -2 \pmod 9$, so $x \equiv 8 \pmod 9$. We see that $(x,y)=(8,32)$ is the smallest solution in positive integers, so $\min(x+y) = 8+32 = \boxed{40}$. □

10. Since $105\times N = 100\times N + 5\times N$, we have

$$\underbrace{111\ldots11}_{\text{105 digits}}00+\underbrace{555\ldots55}_{\text{105 digits}} = 11\underbrace{666\ldots66}_{\text{103 digits}}55.$$

Thus the sum of the digits of the product $105\times N$ is

$$1+1+103\times 6+5+5=\boxed{630}.$$

□

11. With probability $\frac{1}{2^{10}}\binom{10}{i}$, Mark will flip i heads and $10-i$ tails. In this case, the number on the board will be $\frac{2^i}{2^{10-i}} = 2^{2i-10}$. Applying the definition of expectation, where X represents the number on the blackboard after 10 seconds:

$$\begin{aligned}\mathbb{E}(X) &= \sum_{i=0}^{10}\frac{1}{2^{10}}\binom{10}{i}\times 2^{2i-10}\\ &= \frac{1}{2^{20}}\sum_{i=0}^{10}\binom{10}{i}2^{2i}\\ &= \frac{1}{2^{20}}\sum_{i=0}^{10}\binom{10}{i}4^{i}\end{aligned}$$

The sum $\sum_{i=0}^{10}\binom{10}{i}4^i$ is simply the binomial expansion of $(1+4)^{10}=5^{10}$. Hence $\mathbb{E}(X) = \frac{5^{10}}{2^{20}} = \boxed{\frac{5^{10}}{4^{10}}}$.

Alternate solution: For $i=1,\ldots,10$, let X_i be a random variable which equals 2 with probability $\frac{1}{2}$, and $\frac{1}{2}$ with probability $\frac{1}{2}$. Note that $\mathbb{E}(X_i)=\frac{1}{2}\cdot 2+\frac{1}{2}\cdot\frac{1}{2}=\frac{5}{4}$. After 10 seconds, the number on the board is $X=\prod_{i=1}^{10}X_i$. Using the fact that the expected value of the product of independent random variables equals the product of the expected values, we have

$$\mathbb{E}(X) = \mathbb{E}\left(\prod_{i=1}^{10} X_i\right) = \prod_{i=1}^{10} \mathbb{E}(X_i)$$
$$= \boxed{\left(\frac{5}{4}\right)^{10}}$$

□

12. By the triangle inequality, the sum of the lengths of any two sides must be strictly larger than the third side. We see that the only possible side lengths are $(51, 51, 98)$, $(52, 52, 96)$, $(53, 53, 94)$, ..., $(99, 99, 2)$, giving $99 - 51 + 1 = \boxed{49}$ triangles. □

13. Observe that for all integers b, $2\spadesuit b = 2b - 2(2+b) = 2b - 4 - 2b = -4$. Then $1\spadesuit(2\spadesuit(3\spadesuit(4\spadesuit 5))) = 1\spadesuit(-4) = 1(-4) - 2(1-4) = \boxed{2}$. □

14. Using the formula for the number of positive divisors of an integer, we observe that if $d(n) = 4$, then $n = p^3$ or $n = pq$ for distinct primes p and q.

    If either m or $m+1$ is the cube of a prime, we see that $(7, 8)$ or $(8, 9)$ do not give us a solution, but $(26, 27)$ does, so $m = 26$ is a candidate. Otherwise, m and $m+1$ are both the product of two distinct primes. The smallest primes are 2, 3, 5, 7; we quickly see that $2 \times 7 = 14$ and $3 \times 5 = 15$. Hence $m = 14$, so the answer is $1 + 4 = \boxed{5}$. □

15. Let W, L and T represent a win, a loss and a tie respectively. Let A, B, C and D be the four teams who gets 1, 2, 5 and 8 points in total, respectively. Let E be the fifth team.

    Neither A nor B has a win, since both of them have fewer than 3 points, i.e. no wins. So A has $\{T, L, L, L\}$ and B has $\{T, T, L, L\}$. Moreover, the match between A and B must be a tie. Thus A lost all other matches against C, D, E.

    Since C has a win against A and 5 points in total, C has $\{W, T, T, L\}$. Then the match between B and C must be a tie, since neither B nor C has no other wins. Thus B loses its matches against D and E.

    Since D has two wins against A and B, and 8 points in total, D has $\{W, W, T, T\}$, i.e. its other matches against C or E are tie.

    Finally, E has a win against A, B and C, while it has a tie against D. So its total score is $3 \cdot 3 + 1 = \boxed{10}$. □

16. Since $\angle BCD = 90°$, $BC = 13$, and $CD = 11$, we can fix the relative locations of B, C, and D. By the Pythagorean theorem, $BD = \sqrt{13^2 + 11^2} = \sqrt{290}$, so in particular, $AB < BD$. Then A must lie on the circle of radius 17 centered at B:

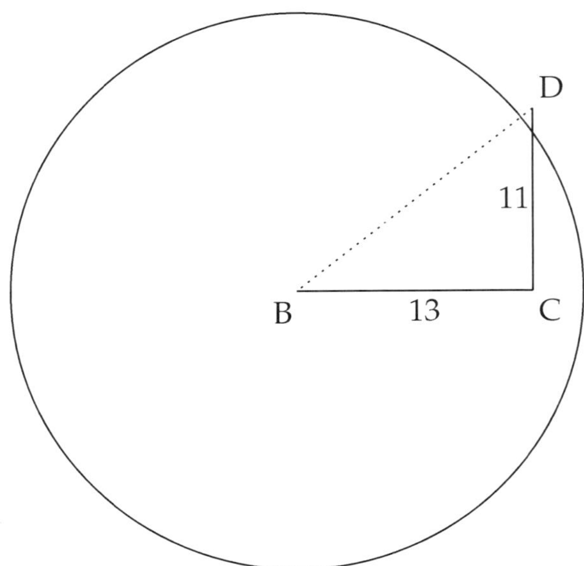

By inspection, there is only one point A which satisfies the conditions that $\angle DAB = 30°$, and that the quadrilateral is not self-intersecting. It is shown in the following diagram:

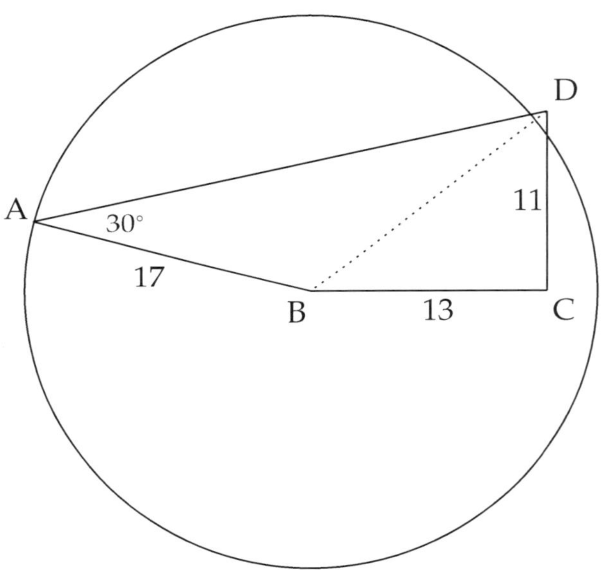

So the number of non-congruent quadrilaterals is $\boxed{1}$. □

17. Recall that we can write the equation of a parabola in vertex form: $y = ax^2 + bx + c = a(x-h)^2 + k$. We show that in fact, $\boxed{\text{All six}}$ quantities can be negative.

    Suppose a, b, c are negative. The x-coordinate of the vertex h is $-\frac{b}{2a}$, which is negative if a and b are negative. The discriminant $\Delta = b^2 - 4ac$ could be negative if a and c are large enough in absolute value.

    As an example, consider $y = -2(x+1)^2 - 1 = -2x^2 - 4x - 3$. Clearly a, b, c, h, and k are negative. The discriminant is $\Delta = (-4)^2 - 4(-2)(-3) = -8 < 0$. Hence in this example, all six of the given quantities are negative. □

18. Because $x+y+z$ is even, exactly one of the primes must be even and the other two are odd, so $x = 2$ and $y+z = 66$.

The second equation gives us $2y+2z+yz = 437$. Because $y+z = 66$, we can substitute to get $2(66) + yz = 437$, or $yz = 437 - 132 = \boxed{305}$. (The solution is $(x,y,z) = (2,5,61)$). $\square$

19. We will represent each positive integer $< 2^{10}$ as a sequence of 10 binary digits, where leading zeros are allowed (e.g., $27 = 0000011011$).

We will use complementary counting, by finding the number of integers whose binary representation *does not* contain two consecutive 1's. Let a_n denote the number of n-digit binary numbers (leading zeros allowed) whose binary representation does not contain two consecutive 1's. Then $a_1 = 2$, $a_2 = 3$, $a_3 = 5$, and in general, $a_n = a_{n-1} + a_{n-2}$ for $n \geq 3$ (why?). The solution is $a_n = F_{n+2}$ where F_n is the nth Fibonacci number with $F_1 = F_2 = 1$, so $a_{10} = F_{12} = 144$. There are 144 numbers less than 2^{10} whose binary representation does not contain two consecutive 1's. The number $2^{10} = 10000000000_2$ does not contain two consecutive 1's and is not counted. Hence the answer is $(2^{10} - 1) - 144 = \boxed{879}$. $\square$

20. See the following diagram:

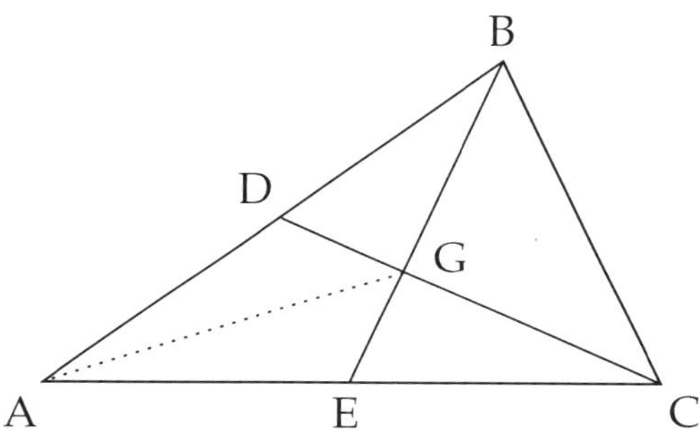

Note that G is the centroid of $\triangle ABC$, and that the centroid divides the medians into segments of ratio $2:1$. Then $BG = 12$ and $CG = 16$. Because $\angle BGC = 90°$, then by the Pythagorean theorem, $BC = 20$.

Extend AG to intersect BC at H. Because AH is a median, we have $BH = HC = 10$, and further, $GH = 10$ since $\triangle BGC$ is right. Then $AG = 2 \times GH = 2 \times 10 = \boxed{20}$. $\square$

21. Cube both sides of the first equation to obtain $a^3b^6 = 1$, which may be written as $a^3(b^3)^2 = 1$. Note that a is necessarily positive, as $a = \frac{1}{b^2} > 0$. However b is not necessarily positive.

Hence we have a system of two equations in terms of a^3 and b^3. Let $a^3 = x$ and $b^3 = y$, where $x > 0$.

$$xy^2 = 1$$
$$x + 3y = 4$$

Substitute x with $4-3y$ into the first equation to obtain $(4-3y)y^2 = 1$, or $3y^3 - 4y^2 + 1 = 0$. Immediately we see that $y = 1$ is a solution; using long or synthetic division, we see the cubic equation factors as $(y-1)(3y^2 - y - 1) = 0$. By the quadratic formula, the other two solutions are $y = \frac{1 \pm \sqrt{13}}{6}$.

If $y = 1$, then $x = 1$, which induces the solution $(a, b) = (1, 1)$. If $y = \frac{1+\sqrt{13}}{6}$, then using $x = 4 - 3y$ we obtain $x = \frac{7-\sqrt{13}}{2}$ which is positive. If $y = \frac{1-\sqrt{13}}{6}$, then $x = \frac{7+\sqrt{13}}{2}$. In all these cases, $(a, b) = (\sqrt[3]{x}, \sqrt[3]{y})$, and $a^3 + b^3$ is simply $x + y$.

The first solution gives $a^3 + b^3 = x + y = 2$. The second solution gives $x + y = \frac{11-\sqrt{13}}{3}$. The third solution gives $x + y = \frac{11+\sqrt{13}}{3}$. The product of all possible values of $a^3 + b^3$ (or $x + y$) is

$$2\left(\frac{11-\sqrt{13}}{3}\right)\left(\frac{11+\sqrt{13}}{3}\right) = \boxed{24}.$$

□

22. Because a and b both have an odd number of divisors, they must be perfect squares. Hence $a = x^2$ and $b = y^2$ for positive integers x and y, and $x^2 + y^2 = 2020$.

    Since $2020 \equiv 0 \pmod 4$, both x and y must be even (since squares of odd integers are $1 \pmod 4$), so let $x = 2k$ and $y = 2m$. Then $k^2 + m^2 = 505$, and we can quickly check that $(k, m) = (21, 8), (12, 19)$ are the solutions (or $(8, 21), (19, 12)$). This gives either $42^2 + 16^2 = 2020$ or $38^2 + 24^2 = 2020$.

    We see that $42^2 = 2^2 \cdot 3^2 \cdot 7^2$ has 27 divisors and $16^2 = 2^8$ has 9 divisors, so $(a, b) = (42^2, 16^2)$ or $(16^2, 42^2)$ does not satisfy the condition. However $38^2 = 2^2 \cdot 19^2$ has 9 divisors and $24^2 = 2^6 \cdot 3^2$ has 21 divisors. Hence $(a, b) = (38^2, 24^2)$, and $a - b = 38^2 - 24^2 = (38 - 24)(38 + 24) = \boxed{868}$. □

23. We use the shoelace formula to find the area of $P_0P_1 \ldots P_n$. As such, it may help to list the points in order:

$$\begin{array}{cc} 1 & 2020 \\ 2 & 1010 \\ 4 & 505 \\ & \square \\ 2^n & \frac{2020}{2^n} \\ 1 & 2020 \end{array}$$

The area of $P_0P_1 \ldots P_n$ is $\frac{1}{2}|4040n + \frac{2020}{2^n} - 1010n - 2020 \times 2^n| = \frac{1}{2}|3030n + \frac{2020}{2^n} - 2020 \times 2^n|$, and we want to find n such that $[P_0P_1 \ldots P_n]$ is greater than 2020^2. That is, we wish to solve the inequality

$$\frac{1}{2}|3030n + \frac{2020}{2^n} - 2020 \times 2^n| > 2020^2$$

for n. Divide both sides by 1010 to obtain

$$\frac{1}{2}|3n + \frac{2}{2^n} - 2 \times 2^n| > 4040$$
$$|3n + \frac{2}{2^n} - 2 \times 2^n| > 8080$$

Note that, for large enough n, the $2 \times 2^n = 2^{n+1}$ term dominates, so we can remove the absolute value signs to obtain

$$2^{n+1} - 3n - \frac{2}{2^n} > 8080$$

A quick check reveals that $n = \boxed{12}$ is the minimum such n. □

24. Let $a = \sqrt[3]{6\sqrt{3}+10} - \sqrt[3]{6\sqrt{3}-10}$. The intuitive first step is to cube both sides to obtain

$$\begin{aligned} a^3 &= (6\sqrt{3}+10) - (6\sqrt{3}-10) \\ &\quad + 3\sqrt[3]{(6\sqrt{3}+10)(6\sqrt{3}-10)^2} - 3\sqrt[3]{(6\sqrt{3}+10)^2(6\sqrt{3}-10)} \\ &= 20 + 3\sqrt[3]{8(6\sqrt{3}-10)} - 3\sqrt[3]{8(6\sqrt{3}+10)} \\ &= 20 + 6\sqrt[3]{6\sqrt{3}-10} - 6\sqrt[3]{6\sqrt{3}+10} \\ &= 20 - 6a \end{aligned}$$

Then $a^3 = 20 - 6a$, or $a^3 + 6a - 20 = 0$. So $x = a$ is a root of the polynomial $x^3 + 6x - 20$, giving $b + c = 6 - 20 = -14$.

But wait! Notice that the polynomial $x^3+6x-20$ has $x = 2$ as a root. In fact, the polynomial $x^3 + 6x - 20$ factors to $(x-2)(x^2+2x+10)$, so its roots are $x = 2$ and $x = \frac{-2\pm\sqrt{2^2-4\cdot 10}}{2} = -1 \pm 3i$. Since $a = \sqrt[3]{6\sqrt{3}+10} - \sqrt[3]{6\sqrt{3}-10}$ is real, it follows that a must simply equal 2. Since $a = 2$ is a root of infinitely many integer polynomials, including $x^3 + 6x - 20$, $x^3 - 8$, $x^3 + 3x - 14$, in which $b + c$ depends on the choice of coefficients, the answer is $\boxed{\text{Cannot be determined}}$. □

25. We will find the probability of the complementary event; that is, we will find the probability that some team either wins all their matches or loses all their matches. Let A denote the event that some team wins all their matches, and let B denote the event that some team loses all their matches.

For all $1 \le i \le 6$, the probability that team #i wins all their matches is $\left(\frac{1}{2}\right)^5 = \frac{1}{32}$. Since the events "team #i wins all their matches" and "team #j wins all their matches" are disjoint for $i \neq j$ (as two teams cannot possibly win all of their matches), we obtain $\mathbb{P}(A) = 6 \times \frac{1}{32} = \frac{3}{16}$. By symmetry, $\mathbb{P}(B) = \frac{3}{16}$.

However, we want the probability that some team wins all their matches *or* some team loses all their matches. We use the fact that for any events A and B, $\mathbb{P}(A \cup B) = \mathbb{P}(A) + \mathbb{P}(B) - \mathbb{P}(A \cap B)$.

To find $\mathbb{P}(A \cap B)$, i.e., the probability that some team wins all matches and some team loses all matches, we observe that there are 6 ways to choose the team (say team #i) to win all matches and 5 ways to choose the team (say team #j, $j \neq i$) to lose all matches. The probability that team #i wins all matches is $\frac{1}{2^5}$; given that this occurs, the probability that team #j loses all matches is $\frac{1}{2^4}$ (not $\frac{1}{2^5}$, as the match between teams #i and #j was already decided). The remaining matches in the tournament are arbitrary, so we obtain $\mathbb{P}(A \cap B) = 6 \times 5 \times \frac{1}{2^9} = \frac{15}{256}$.

Using the inclusion-exclusion rule above, we obtain $\mathbb{P}(A \cup B) = \frac{3}{16} + \frac{3}{16} - \frac{15}{256} = \frac{81}{256}$. However this represents the probability of the complementary event, so the probability that every team wins and loses at least one match is $1 - \frac{81}{256} = \boxed{\frac{175}{256}}$. □

1. A total of $\binom{8}{2} = 28$ games were played, so the sum of the eight players' scores must be 28. Then the eighth player scored $28 - (0 + 1 + 2.5 + 3 + 3.5 + 5 + 7) = \boxed{6}$ points.

   To show that such a tournament is possible, suppose the players were ranked #1, #2, ..., #8, where #1 scored 7 points, #2 scored 6 points, and so on. A possible tournament outcome is as follows: #1 beat everyone, #2 beat #3 through #8, #3 beat #4 through #8, #4 tied #5 and beat #6-8, #5 tied #4 and #6 and beat #7 and #8, #6 tied #5 and beat #7 and #8, #7 beat #8, and #8 lost to everyone. □

2. Since $B \times C$ has units digit 1, we see that $B \times C$ must equal 21 (11, 31, 41, 61, 71 do not work as these are prime; 51 does not work as $51 = 3 \times 17$, 91 does not work as $91 = 7 \times 13$, and 1 and 81 do not work as $B = C$). Then B and C must be 3 and 7, in some order.

   Suppose $B = 3$ and $C = 7$. We want $A3 \times 37$ to equal the number $7A71$. However, $37 \times 100 = 3700 < 7A71$, so no two-digit number satisfies this property.

   Now, suppose $B = 7$ and $C = 3$. We want $A7 \times 73 = 3A31$. Testing, we see $A = 4$ works as $47 \times 73 = 3431$, so the answer is $\boxed{4}$. □

3. The units digit must be 5 or 0; because the digits are odd, the units digit must be 5. Each of the remaining three digits can be 1, 3, 5, 7, or 9, giving $5^3 = \boxed{125}$ numbers. □

4. We claim that P is the intersection of the diagonals of rhombus $ABCD$. By the triangle inequality, we have $PA + PC \geq AC$, and $PB + PD \geq BD$, so adding these inequalities yields $PA + PB + PC + PD \geq AC + BD$.

   Since $\angle ABC = 60^\circ$, the rhombus consists of two equilateral triangles $\triangle ABC$ and $\triangle ACD$, of side 6. Then $PA + PC = 6$, and $PB = PD = 3\sqrt{3}$, so $PA + PB + PC + PD = 6 + 2(3\sqrt{3}) = \boxed{6 + 6\sqrt{3}}$. □

5. Let g be the number of games he has bowled, *before* today. The sum of his scores is $180g$. As $193 + 207 + 200 = 600$, we can set up an equation to solve for g: $\frac{180g+600}{g+3} = 182$. Multiply by $g + 3$ to obtain $180g + 600 = 182(g + 3) = 182g + 546 \implies 2g = 54 \implies g = 27$. Including today's games, he has bowled $g + 3 = 27 + 3 = \boxed{30}$ games. □

6. We observe the following property about every number in the sequence: each number in the sequence must leave a remainder of 1 when divided by 3 (to see why, we have that $x \equiv 1 \pmod 3 \implies x + 6 \equiv 1 \pmod 3$, and $x \equiv 1 \pmod 3 \implies 4x \equiv 4 \equiv 1 \pmod 3$).

   Out of all the answer choices, only $\boxed{335}$ does not leave a remainder of 1 when divided by 3. To see that all other answer choices could appear in the sequence, two possible sequences are 1, 7, 13, 19, ... (which includes all numbers congruent to $1 \pmod 6$), or 1, 4, 10, 16, 22, ... (which includes all numbers congruent to $4 \pmod 6$). □

7. The probability is 1 minus the probability of flipping all heads, which is $1 - \left(\frac{2}{3}\right)^3 = \boxed{\frac{19}{27}}$. □

8. We can solve this problem using coordinates. Let $A(0, 0)$, $B(4, 0)$, $C(4, 4)$, and $D(0, 4)$. Let I be the intersection of AF and DE. The line AF is determined by the equation $y = \frac{1}{2}x$, and the line DE is determined by the equation $y = -2x + 4$. The intersection of these two lines is $\left(\frac{8}{5}, \frac{4}{5}\right)$.

Similarly, let J be the intersection of BG and CH. We see that the coordinates of J are $\left(\frac{12}{5}, \frac{16}{5}\right)$.

Then using the distance formula/Pythagorean theorem, we have $IJ^2 = \left(\frac{4}{5}\right)^2 + \left(\frac{12}{5}\right)^2 = \frac{160}{25} = \frac{32}{5}$. The area of a square with diagonal d is $d^2/2$, so the area is $\frac{32}{10} = \boxed{3.2}$. □

9. Because a polynomial always has three roots (including multiplicity), the cubic polynomial must factor into the form $(x-a)^2(x-b)$, where a and b are the integer roots. Expanding, we obtain $(x^2-2ax+a^2)(x-b) = x^3-(2a+b)x^2+(a^2-2ab)x-a^2b = x^3+9x^2+24x+16$.

   Then $a^2b = -16$; since a and b are integers, we know that b must be negative (otherwise $a^2b > 0$). Hence the only possibilities for b are -1, -2, and -4, implying $a^2 = 16$, 4, 1, respectively. The only possibility that works is $(a, b) = (-4, -1)$, in which the polynomial factors as $(x+4)^2(x+1)$, so $a+b = -4-1 = \boxed{-5}$. □

10. Let $d(k)$ denote the number of divisors of k. We observe that $\gcd(n, 100) = 1$, since n ends in 99 and cannot be divisible by 2 or 5. Using the fact that $d()$ is multiplicative, we have $d(100n) = d(100)d(n) = 6d(100)$. Since $100 = 2^2 \times 5^2$, we have $d(100) = (2+1)(2+1) = 9$, so the answer is $6 \times 9 = \boxed{54}$. □

11. Consider the shape induced by the four digits of the PIN, which must be one of the following tetrominos (up to rotation and reflection):

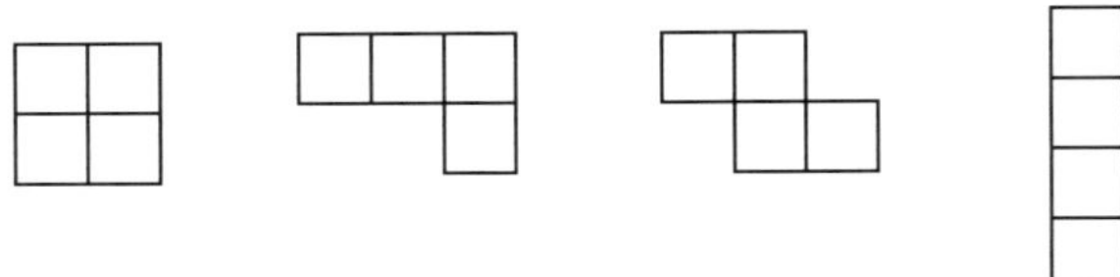

   For example, the PINs 1254 and 4587 trace out a square, and the PIN 1256 traces out a Z tetromino. For brevity, we will label these tetrominos T_1, T_2, T_3, T_4 from left to right.

   For the case T_1, there are four possible squares on the keypad. For each square (e.g., the square containing the numbers 1, 2, 4, 5), there are 8 possible PINs that trace that square (4 ways to choose the first digit, 2 ways to choose the direction traced out by the PIN), giving $4 \times 8 = 32$ possible PINs.

   For the case T_2, there are 18 L-tetrominos on the keypad (e.g., 1-2-3-6, 1-4-5-6, 3-6-5-4, etc.), and each L-tetromino gives 2 possible PINs, or $18 \times 2 = 36$ possible PINs.

   For the case T_3, there are 10 S-tetrominos, and each one gives 2 possible PINs, or $10 \times 2 = 20$ possible PINs.

   For the case T_4, there is only one I-tetromino (2-5-8-0), giving $1 \times 2 = 2$ possible PINs.

   Altogether we have $32 + 36 + 20 + 2 = \boxed{90}$ different 4-digit PINs. □

12. By the angle bisector theorem, $AD : DC = 5 : 13$, so $AD = \frac{10}{3}$ and $DC = \frac{26}{3}$.

   Now consider right triangle ABD. By the Pythagorean theorem, the length of hypotenuse BD is $\frac{5\sqrt{13}}{3}$. As M is the midpoint of hypotenuse BD, it is also the circumcenter of $\triangle ABD$, so we see that $BM = MD = MA$ and $\triangle AMD$ is isosceles. Then $m\angle DAM = m\angle ADM$, and $\sin m\angle DAM = \sin m\angle ADM = \frac{5}{5\sqrt{13}/3} = \frac{3}{\sqrt{13}} = \boxed{\frac{3\sqrt{13}}{13}}$. □

13. Let $a = 2^{2020}$, so that the desired expression equals 2^a. Since $2^a = 4^k = (2^2)^k = 2^{2k}$, we have $a = 2k \implies k = \frac{a}{2} = \boxed{2^{2019}}$. □

14. The sequence is identical to counting in base 5, except that 0, 1, 2, 3, 4 are replaced with 0, 2, 4, 6, 8. The answer is the value of 1010_5, which is $5^3 + 5 = \boxed{130}$. □

15. Observe that '20' must be parsed as T, then the following 8 and 9 must be parsed as HI. Further, the subsequence '35' occurs twice, so both of these 5's must be parsed as E's. Hence we can divide up the task into finding the number of ways to parse the strings $S_1 = 131$, $S_2 = 19123$, and $S_3 = 191513$ separately. By the multiplication principle (rule of product), the number of possible ways is simply the product of the numbers of ways to parse S_1, S_2, and S_3.

    First, we see that $S_1 = 131$ can be parsed two ways (ACA, MA).

    Second, we count the number of ways to parse $S_2 = 19123$. Notice that a split must occur after the '9'. Thus we can parse the first two characters '19' two ways (AI or S), then parse the last three characters '123' three ways (ABC, LC, AW). This gives $2 \times 3 = 6$ ways to parse S_2.

    Similarly for $S_3 = 191513$, a split must occur after the 9, and a split must occur after the 5. This gives $2 \times 2 \times 2 = 8$ ways to parse S_2.

    Then the number of possible strings is $2 \times 6 \times 8 = \boxed{96}$. □

16. *Solution 1 (Heron's, $A = \frac{abc}{4R}$)* We have the following diagram:

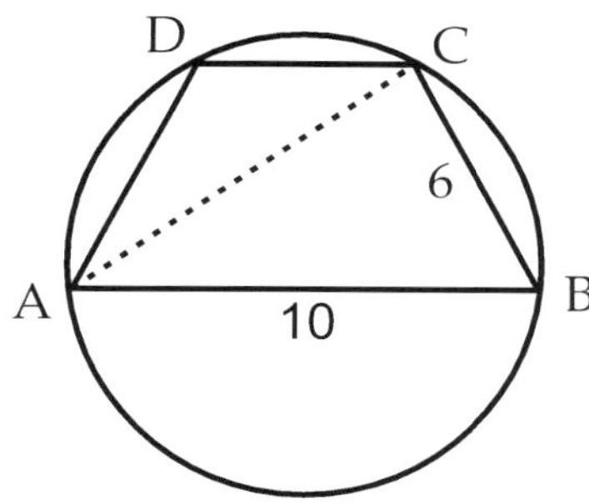

We find the length of AC. There are multiple ways to find AC, but a fast way uses Ptolemy's theorem, which gives us

$$10 \cdot 6 + 6 \cdot 6 = AC^2 \implies AC = 4\sqrt{6}$$

Next, we use Heron's formula to find the area of $\triangle ABC$. The semiperimeter equals $\frac{10+6+4\sqrt{6}}{2} = 8 + 2\sqrt{6}$, so Heron's gives us

$$[\triangle ABC] = \sqrt{(8+2\sqrt{6})(8-2\sqrt{6})(2+2\sqrt{6})(-2+2\sqrt{6})}$$

The RHS simplifies nicely by difference of squares, giving us $[\triangle ABC] = \sqrt{40 \cdot 20} = 20\sqrt{2}$. Lastly, applying the area formula $A = \frac{abc}{4R}$ on $\triangle ABC$, we have

$$20\sqrt{2} = \frac{10 \cdot 6 \cdot 4\sqrt{6}}{4 \cdot R} = \frac{60\sqrt{6}}{R} \implies R = 3\sqrt{3}$$

The area of the circle equals $\pi r^2 = \pi(3\sqrt{3})^2 = \boxed{27\pi}$.

Solution 2 (Algebra) Let O denote the center of the circle. Notice that O is contained inside $\triangle ABC$ as $\triangle ABC$ is acute (this can be checked as $(4\sqrt{6})^2 + 6^2 > 10^2$).

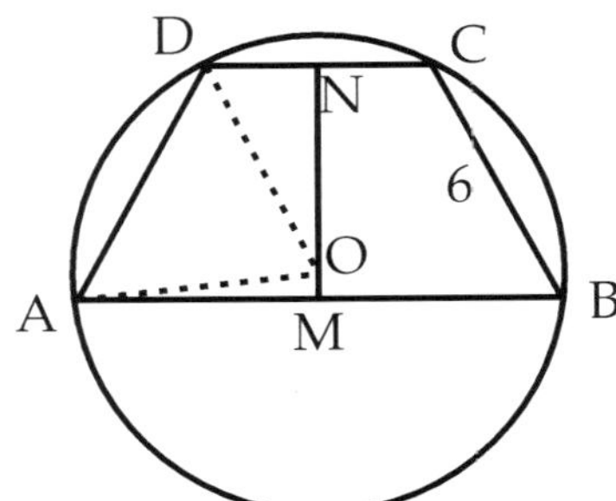

Let M and N be the midpoints of AB and CD, respectively, so that $AM = 5$ and $DN = 3$. Then we have $OM = \sqrt{R^2 - 25}$ and $ON = \sqrt{R^2 - 9}$ by the Pythagorean theorem. Further, it is not too hard to see that $MN = 4\sqrt{2}$, also by the Pythagorean theorem. Hence we have

$$\sqrt{R^2 - 25} + \sqrt{R^2 - 9} = 4\sqrt{2}$$

Let $x = R^2$, and square both sides to get

$$\begin{aligned}(x - 25) + (x - 9) + 2\sqrt{(x - 25)(x - 9)} &= 32\\ 2\sqrt{x^2 - 34x + 225} &= -2x + 66\\ \sqrt{x^2 - 34x + 225} &= -x + 33\end{aligned}$$

Squaring again gives

$$\begin{aligned}x^2 - 34x + 225 &= x^2 - 66x + 1089\\ 32x &= 864\\ x &= 27\end{aligned}$$

Hence $R = 3\sqrt{3}$ and the area of the circle is $x\pi = 27\pi$ as before.

Solution 3 (Parameshvara's circumradius formula): Let $a = 10$, $b = c = d = 6$. The semiperimeter is $s = 14$. The circumradius R equals

$$\begin{aligned}R &= \frac{1}{4}\sqrt{\frac{(ab + cd)(ac + bd)(ad + bc)}{(s - a)(s - b)(s - c)(s - d)}}\\ &= \frac{1}{4}\sqrt{\frac{(60 + 36)(60 + 36)(60 + 36)}{(14 - 10)(14 - 6)(14 - 6)(14 - 6)}}\\ &= 3\sqrt{3}\end{aligned}$$

□

17. By considering the graph of $\ln|\tan x|$, we see that on the interval $(0, \pi)$, there are exactly two values x such that $\ln|\tan x| = 2020$ and two values x such that $\ln|\tan x| = -2020$. So there are exactly four values x in $(0, \pi)$ such that $f(x) = 2020$. As f is periodic with period π, the answer is $4 \times 2020 = \boxed{8080}$. □

18. Observe that $792 = 8 \times 9 \times 11$. Using the divisibility rule for 9, remaining digit is 8, and the sum of the digits is $1 + 2 + 3 + 4 + 8 = 18$.

    Using the divisibility rule for 11, the difference between even-positioned and odd-positioned digits must be $0 \pmod{11}$, so the only possibility is that the even-positioned digits are 8, 1 (in some order), and the odd-positioned digits are 2, 3, 4 (in some order).

    For the case _8_1_, since the number is divisible by 8, the units digit must be 2, and the only possibility is 48312. For the case _1_8_, the only possibility is 21384. In either case, the hundreds digit is $\boxed{3}$. □

19. We can do casework on the number of \$5 bills used. Notice that if the cashier uses k \$5 bills, then the number of \$2 bills used must be so that the total value of the \$2 and \$5 bills is at least \$96.

    - If the cashier uses 20 \$5 bills, then he must use 0 \$2 and 0 \$1 bills, or 1 way.
    - If the cashier uses 19 \$5 bills, then he can use 1 or 2 \$2 bills, or 2 ways.
    - If the cashier uses 18 \$5 bills, then he can use 3, 4, or 5 \$2 bills, or 3 ways.
    - If the cashier uses 17 \$5 bills, then he can use 6 or 7 \$2 bills, or 2 ways.

    Observe that if the cashier uses $2k$ \$5 bills (where $0 \le k \le 9$), then there are 3 ways, and if the cashier uses $2k+1$ \$5 bills ($0 \le k \le 9$), then there are 2 ways. Altogether, there are $1 + 3(10) + 2(10) = \boxed{51}$ ways.

    Alternate solution: Consider the generating function $G(x) = (1+x+x^2+x^3+x^4)(1+x^2+x^4+\ldots)(1+x^5+x^{10}+\ldots)$. The coefficient of x^{100} represents the number of ways to make \$100. Expanding, we have

$$\begin{aligned}
G(x) &= (1+x+x^2+x^3+x^4)(1+x^2+x^4+\ldots)(1+x^5+x^{10}+\ldots) \\
&= \frac{1-x^5}{1-x}\cdot\frac{1}{1-x^2}\cdot\frac{1}{1-x^5} \\
&= \frac{1}{1-x}\cdot\frac{1}{1-x^2} \\
&= (1+x+x^2+\ldots)(1+x^2+x^4+\ldots)
\end{aligned}$$

    Here we can see that the x^{100} coefficient is $\boxed{51}$. □

20. The slant height of the cone is $\sqrt{3^2+4^2} = 5$ cm.

    Consider a cross-section of the cone and sphere, which will contain a 5-5-6 triangle and its inscribed circle. The radius r of the sphere equals the inradius of a 5-5-6 triangle. Using the area formula $A = rs$, we see that $12 = 8r \implies r = \frac{3}{2}$. The volume of the sphere is $\frac{4}{3}\pi r^3 = \frac{4}{3}\pi\left(\frac{3}{2}\right)^3 = \boxed{\frac{9}{2}\pi}$. □

21. Let $y = x + \frac{1}{x}$. Then $y^2 = x^2 + \frac{1}{x^2} + 2$, so $2y^2 = 2x^2 + \frac{2}{x^2} + 4 \implies 2y^2 - 4 = 2x^2 + \frac{2}{x^2}$. Then the objective is equivalent to minimizing $2y^2 + y - 4$. Further, it can be easily be shown that $y \geq 2$. The quadratic $2y^2 + y - 4$ is increasing on the interval $[2, \infty)$, so the minimum value is $2 \times 2^2 + 2 - 4 = \boxed{6}$, which occurs when $y = 2$ (or $x = 1$).

Alternate solution: By the AM-GM inequality, for all positive x, $\frac{x+\frac{1}{x}+x^2+x^2+\frac{1}{x^2}+\frac{1}{x^2}}{6} \geq \sqrt[6]{1} = 1$, so the minimum is at least 6. Equality is achieved when $x = 1$.

Note that we can apply AM-GM on the expression directly to obtain $\frac{x+\frac{1}{x}+2x^2+\frac{2}{x^2}}{4} \geq 1 \implies$ $x + \frac{1}{x} + 2x^2 + \dfrac{2}{x^2} \geq 4$. But 4 is not the minimum, as equality is not attained. □

22. We observe that $f^n(x)$ is an even function for all n. This can be shown inductively; $f^1(x)$ is clearly even, and if $f^n(x)$ is even, then $f^{n+1}(-x) = f^1(f^n(-x)) = f^1(f^n(x)) = f^{n+1}(x)$, so f^{n+1} is also even. Then $f(x) = 0$ if and only if $f(-x) = 0$, so $x_0 = -x_1$. Hence it suffices to find x_1.

Let a_n denote the largest real root of the polynomial $f^n(x)$, so that $x_1 = a_{2020}$. We see that $a_1 = \sqrt{20}$ (as $x^2 - 20 = 0 \implies x = \pm\sqrt{20}$), $a_2 = \sqrt{20 + \sqrt{20}}$, $a_3 = \sqrt{20 + \sqrt{20 + \sqrt{20}}}$, and so on. More generally, (a_n) converges to the value of $\sqrt{20 + \sqrt{20 + \sqrt{20 + \ldots}}}$; setting this infinitely nested radical equal to a, we have $a = \sqrt{20 + a} \implies a^2 - a - 20 = 0$, or $a = -4, 5$. As $a > 0$, we take the positive solution $a = 5$.

As $x_1 = a_{2020} < 5$, we see that $x_1 = 4.999\ldots < 5$, and $x_0 = -x_1$, so $\lfloor x_0^2 + x_1^2 \rfloor = \boxed{49}$. □

23. There are 2^{10} possible outcomes. We find the number of outcomes containing no three consecutive H's or T's using recursion.

Let a_n equal the number of outcomes with n coin flips containing no three consecutive H's or T's. We have $a_1 = 2$, $a_2 = 4$, and $a_3 = 2^3 - 2 = 6$. In order to compute a_n, we can append an H or T to any valid length-$(n-1)$ outcome, giving $2a_{n-1}$ ways. However, if the $(n-2)$th and $(n-1)$th coin flip are H (or T), and we appended an H (or T), we obtain an invalid outcome as three consecutive heads (or tails) resulted. Hence we must subtract the number of outcomes for which the last three flips are the same. It can be checked that this gives a_{n-3} outcomes.

Hence our recurrence is $a_n = 2a_{n-1} - a_{n-3}$ for $n \geq 4$. Computing, we have $a_4 = 10$, $a_5 = 16$, $a_6 = 26$, $a_7 = 42$, $a_8 = 68$, $a_9 = 110$, and $a_{10} = 178$. The desired probability is $\frac{178}{2^{10}} = \boxed{\dfrac{89}{512}}$. (Note: you may observe that it is also true that $a_n = a_{n-1} + a_{n-2}$) □

24. We will write $Q(x)$ in factored form, then use difference of squares:

$$\begin{aligned}Q(x) &= (x - z_1^2)(x - z_2^2)(x - z_3^2) \\ &= (x - z_1)(x + z_1)(x - z_2)(x + z_2)(x - z_3)(x + z_3) \\ &= \prod_{i=1}^{3}(x - z_i)\prod_{i=1}^{3}(x + z_i) \\ &= P(x)(-1)^3P(-x) \\ &= -P(x)P(-x)\end{aligned}$$

To find $a + b + c$, we will find $Q(1)$, then $a + b + c = Q(1) - 1$.

$$\begin{aligned}Q(1) &= -P(1)P(-1) \\ &= -2023 \times 2017 = -(2020^2 - 9) = -4080391\end{aligned}$$

Then $a + b + c = Q(1) - 1 = \boxed{-4080392}$. □

25. Note that the probability of flipping n heads is $\frac{\binom{10}{n}}{2^{10}}$. If X denotes the number of dollars the player wins, then by definition of expectation, we have

$$\mathbb{E}(X) = 0^2 \cdot \frac{\binom{10}{0}}{2^{10}} + 1^2 \cdot \frac{\binom{10}{1}}{2^{10}} + \ldots + 10^2 \cdot \frac{\binom{10}{10}}{2^{10}} = \frac{1}{2^{10}}\sum_{n=0}^{10} n^2\binom{10}{n}$$

Note we can ignore $n = 0$ as $0^2\binom{10}{0} = 0$. To evaluate $\sum_{n=1}^{10} n^2\binom{10}{n}$, we consider the following combinatorial interpretation. Consider a room containing 10 people, in which we want to select a committee of size $n \geq 1$. From this committee, we will select one person as president and one person as vice president, where the same person can fulfill both roles. Summing from $n = 1$ to $n = 10$, the number of ways to do this is $\sum_{n=1}^{10} n^2\binom{10}{n}$. Alternately, we can select the president and vice president first, then select the rest of the committee freely.

If the president and vice president are the same person, then there are 10 ways to choose the person to fulfill both roles, then $2^9 = 512$ ways to select the rest of the committee. If the president and vice president are different, then there are $10 \times 9 = 90$ ways to fulfill these roles, then $2^8 = 256$ ways to select the rest of the committee. Hence we obtain

$$\sum_{n=0}^{10} n^2\binom{10}{n} = 10 \times 2^9 + 90 \times 2^8 = 28160$$

Substituting back into our formula for $\mathbb{E}(X)$, we have

$$\mathbb{E}(X) = \frac{1}{2^{10}}(10 \times 2^9 + 90 \times 2^8) = \frac{10}{2} + \frac{90}{4} = 27.5$$

Therefore the mathematician should charge $\boxed{\$27.50}$ for the game to be fair.

Alternate solution: Let X be a random variable indicating the number of heads flipped (out of 10 coins), so that $\mathbb{E}(X) = 5$. The desired answer is $\mathbb{E}(X^2)$. We will use the fact that

$\text{Var}(X) = \mathbb{E}(X^2) - \mathbb{E}(X)^2$. To find $\text{Var}(X)$, we use the facts that the variance of a 0-1 variable with probability p of success is $p(1-p)$, and that the variance of a sum of independent random variables is the sum of the variances. Hence $\text{Var}(X) = 10 \times \frac{1}{2}\left(1 - \frac{1}{2}\right) = \frac{5}{2}$, and $\mathbb{E}(X^2) = \text{Var}(X) + \mathbb{E}(X)^2 = \frac{5}{2} + 25 = \boxed{27.5}$. □

1. Suppose Jack traveled D miles from San Diego to Los Angeles. Then he traveled $1.5D$ miles on the return trip. He used a total of $\frac{D}{40} + \frac{1.5D}{20} = \frac{4D}{40} = \frac{D}{10}$ gallons. He traveled $D + 1.5D = 2.5D$ miles, so his overall fuel efficiency is $\frac{2.5D}{D/10} = \boxed{25.0}$ miles per gallon. □

2. The first five terms are 1, 1, 2, 6, and 42. The sixth term is $1^2 + 1^2 + 2^2 + 6^2 + 42^2 = 1806$. Note that we only care about the last two digits. Thus we can establish that the seventh term of the sequence is congruent to $1^2 + 1^2 + 2^2 + 6^2 + 42^2 + 6^2 \equiv 42 \pmod{100}$. Hence it can be shown that the last two digits of the sequence alternate: 01, 01, 02, 06, 42, 06, 42, 06, ..., so the tenth term ends in $\boxed{06}$. □

3. Either all three spins are even, or two spins are odd and the third is even. The first case occurs with probability $\left(\frac{2}{5}\right)^3 = \frac{8}{125}$. The second case occurs with probability $3 \times \frac{2}{5} \times \left(\frac{3}{5}\right)^2 = \frac{54}{125}$, where the multiplication by 3 accounts for the number of ways to permute two O's and one E. Altogether, the probability is $\frac{8}{125} + \frac{54}{125} = \boxed{\frac{62}{125}}$. □

4. We have $BD = 5$, and by the angle bisector theorem, $DP : PB = 3 : 4$ and $DQ : QB = 4 : 3$. It follows that $DP = QB = \frac{3}{7} \cdot 5 = \frac{15}{7}$ and $PQ = \frac{5}{7}$, or more simply, $PQ = \frac{1}{7}BD$.

   Hence $[\triangle APQ] = \frac{1}{7}[\triangle ABD] = \frac{6}{7}$, and similarly $[\triangle CPQ] = \frac{6}{7}$, where $[R]$ denotes the region of R. The area of quadrilateral $APCQ$ is $\frac{6}{7} + \frac{6}{7} = \boxed{\frac{12}{7}}$. □

5. We simplify the left-hand side using difference of squares:

$$\begin{aligned}(n + 3^{2019})^2 - (n - 3^{2019})^2 &= [(n + 3^{2019}) + (n - 3^{2019})][(n + 3^{2019}) - (n - 3^{2019})] \\ &= (2n)(2 \times 3^{2019}) \\ &= 3^{2019} + 3^{2020}\end{aligned}$$

   Divide both sides of the equation $2n(2 \times 3^{2019}) = 3^{2019} + 3^{2020}$ by 3^{2019} to obtain $4n = 1 + 3 = 4$, or $n = \boxed{1}$. □

6. Since $\frac{n^2+4n}{n+4} = n \in \mathbb{Z}$, we see that $\frac{n^2}{n+4}$ is an integer if and only if $\frac{4n}{n+4}$ is an integer. By similar reasoning, $\frac{4n+16}{n+4}$ is an integer, so the desired expression is an integer if and only if $\frac{16}{n+4}$ is an integer.

   Then $n + 4$ can equal any divisor (positive or negative) of 16, namely $n + 4 = \pm 1, \pm 2, \pm 4, \pm 8$, or ± 16. This gives $\boxed{10}$ values for n. □

7. We solve by complementary counting, by counting the number of ways to divide students so that some team does not have a girl. There are $\binom{10}{5} = 252$ ways to choose a subset of five to form a team, but we overcounted by a factor of 2, since choosing a set $S \subset A = \{1, 2, \ldots, 10\}$ for the first team is equivalent to choosing $A \quad S$. Hence there are 126 distinguishable ways to divide the 10 students into two teams.

   The number of "bad" outcomes (where some team does not have a girl) can easily be shown to be 6. This is because there are 6 ways to divide the students into teams where one team has all four girls and one of the six boys.

Hence the total number of distinguishable outcomes is $126 - 6 = \boxed{120}$.

□

8. Consider a net (2-D visualization of the faces) of the box. The ant must crawl in a straight line along two of the faces, as shown by the dotted paths:

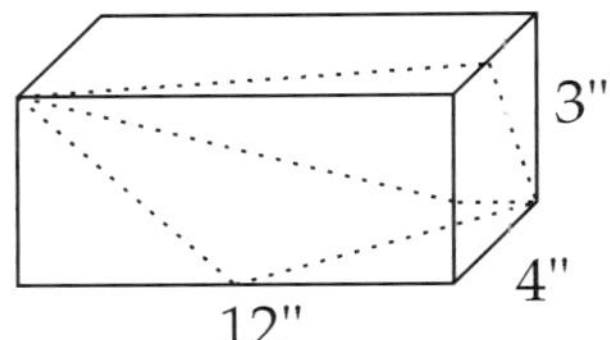

Using the Pythagorean theorem, we see that the lengths of these three possible paths (in inches) are $\sqrt{16^2+3^2}$, $\sqrt{15^2+4^2}$, and $\sqrt{12^2+7^2}$. Of these, $\sqrt{12^2+7^2} = \sqrt{193}$ is the smallest, which is approximately $\boxed{13.9 \text{ in}}$. □

9. We have $(10A + B) + (10B + C) = (10C + A) = 100A + 10B + C$, or $11(A + B + C) = 100A + 10B + C$. This simplifies to $89A = B + 10C$, and since B and C are digits, the only possibility is $A = 1$, $B = 9$, and $C = 8$. It can be checked that $19 + 98 + 81 = 198$. So $A + B + C = \boxed{18}$. □

10. One of a, b, c is an even prime (namely 2), otherwise $a + b + c$ will be odd. Since $a - c$ is even, both a and c are odd and so $b = 2$. Thus the equations $a - c = 8300$ and $a + c = 9702$ give $a = 9001$ and $c = 701$. So $a + 2b = 9001 + 2 \times 2 = \boxed{9005}$. □

11. Suppose the seats are numbered 1, 2, ..., 10. We first find the number of ways to select three non-consecutive seats (or numbers) a, b, $c \in \{1, 2, 3, \ldots, 10\}$. Let $b = a + d_1$ where $d_1 \geq 2$, and let $c = b + d_2$ where $d_2 \geq 2$. We can equivalently count the number of positive integer solutions to $a + d_1 + d_2 \leq 10$, where $a \geq 1$ and $d_1, d_2 \geq 2$.

    Let $a' = a - 1$, $d_1' = d_1 - 2$, and $d_2' = d_2 - 2$. Then this equivalently counts the number of non-negative integer solutions (a', d_1', d_2') to the inequality $a' + d_1' + d_2' \leq 5$. Add a dummy variable $s := 5 - a' - d_1' - d_2'$. This is equivalent to finding the number of non-negative integer solutions (a', d_1', d_2', s) to the equation $a' + d_1' + d_2' + s = 5$. By stars and bars, the number of solutions is $\binom{8}{3} = 56$.

    We multiply by $3! = 6$, since there are 6 ways for the three people to seat themselves, once the seats have been chosen. The number of ways is $56 \times 6 = \boxed{336}$. □

12. Note that a triangle is acute if and only if the sum of the squares of the lengths of any two sides is strictly greater than the length of the third side. This implies the following inequalities must hold:

$$7^2 + 14^2 > c^2$$
$$7^2 + c^2 > 14^2$$
$$c^2 + 14^2 > 7^2$$

The third inequality is always true so we will focus on the first two inequalities. These are equivalent to $c^2 < 245$ and $c^2 > 147$ respectively. The integers c which satisfy both inequalities simultaneously are $c = 13, 14, 15$, or $\boxed{3}$ values. □

13. Let $a = x - 2020$ and $b = y - 2019$. We have $\frac{ab}{a^2+b^2} = \frac{1}{2} \iff a^2 + b^2 = 2ab \iff (a-b)^2 = 0 \iff a = b$. Hence $x - 2020 = y - 2019$, so $x - y = \boxed{1}$. □

14. Note that $s(n)$ is at least 0 and at most 27 if n is a 3-digit number, or at most 36 if n is a 4-digit number. Clearly, we see n should be a 3-digit number (otherwise $n \geq 1000$, in which $s(n) + 3n \geq 3000 > 2019$), so we obtain $3n = 2019 - s(n) \geq 1982$, or $n \geq 661$. Since $s(n) \geq 0$, we have the simple upper bound $3n \leq 2019$, or $n \leq 673$.

    Hence the candidate values for n we need to check are 661, 662, ..., 673. To speed this up, observe that $s(n) \equiv n \pmod 9$, so $s(n) + 3n \equiv 4n \equiv 2019 \equiv 3 \pmod 9$. This implies that $4n \equiv 3 \pmod 9$, or $n \equiv 3 \pmod 9$. The only candidate value for n which is $3 \pmod 9$ is 669 which does not work. Hence there are $\boxed{0}$ integers n satisfying this property. □

15. We can find the probability that Max misses exactly 16 points. This happens if and only if Max incorrectly answers two 8-point problems, but answers everything else correctly. Out of 5^{32} total outcomes, there are $\binom{8}{2} = 28$ ways for Max to answer two problems incorrectly, then $4 \times 4 = 16$ ways for Max to incorrectly answer those two problems. The number of outcomes for which Max scores exactly 200 is $28 \times 16 = \boxed{448}$. □

16. There are 40 segments total in the figure. We see that at most 16 segments can be removed; this uses a basic fact from graph theory that a tree on $|V|$ vertices contains exactly $|V| - 1$ edges, so removing more than 16 edges will disconnect the graph (so that it will be impossible to travel between some pair of vertices).

    We see that removing $\boxed{16}$ segments is possible:

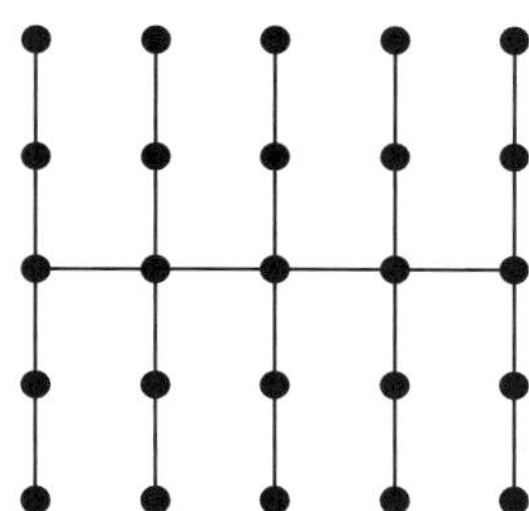

□

17. We write $r_1^2 + r_2^2 + r_3^2 = (r_1 + r_2 + r_3)^2 - 2(r_1r_2 + r_1r_3 + r_2r_3)$. By Vieta's formulas, $r_1 + r_2 + r_3 = 0$ and $r_1r_2 + r_1r_3 + r_2r_3 = 2020$. Substituting, we obtain $0^2 - 2(2020) = \boxed{-4040}$. □

18. We use the fact that a positive integer N is congruent to the sum of its digits modulo 9. Hence $N \equiv 1 + 2 + 3 + \dots + 3 + 9 + 4 + 0 + 4 + 1 \pmod 9$. We can use this fact again to establish $N \equiv 1 + 2 + 3 + \dots + 39 + 40 + 41 \pmod 9$, or $N \equiv 41 \times 21 \equiv \boxed{6} \pmod 9$.

    Alternate solution: We can find the sum of digits of N directly by grouping the digits. The sum of digits from 1 to 9 is $1 + 2 + \dots + 9 = 45$. The sum of the digits of the numbers 10, 11, ..., 19 is $10 + (1 + 2 + \dots + 9) = 55$. Similarly, the sums of the digits of the numbers from 20-29 and 30-39 are 65 and 75, respectively. Finally, $4 + 0 + 4 + 1 = 9$. Then the sum of the digits of N is $45 + 55 + 65 + 75 + 9 = 249$, so the remainder when N is divided by 9 is $249 \pmod 9 = \boxed{6}$. □

19. Consider the five circles in the topmost row or the leftmost column. There are 3^5 ways to assign the numbers 1, 2, or 3 to these circles. This uniquely determines the remaining four circles, so the number of ways is $\boxed{3^5}$. □

20. Consider the following diagram:

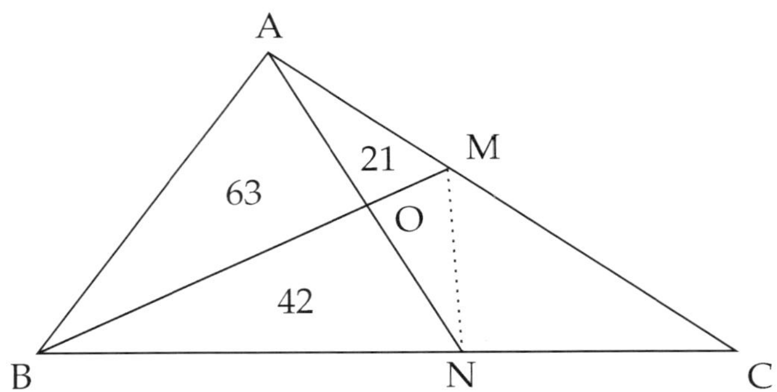

Let $[X]$ denote the area of X. We see that $AO : ON = 3 : 2$ since $[\triangle AOB] : [\triangle OBN] = 3 : 2$, so $[\triangle MON] = 14$. Let $[\triangle MNC] = a$. Then $\frac{a}{14+42} = \frac{a+14+21}{42+63}$, or $\frac{a}{56} = \frac{a+35}{105}$. Solving for a, we get $a = \boxed{40}$. □

21. Let x and y be real numbers with $x+y = n$ and $x^2+y^2 = n+19$. We observe the following inequality:

$$2x^2 + 2y^2 \geq (x+y)^2$$

This can be shown by noting that $x^2+y^2 \geq 2xy$, then adding x^2+y^2 to both sides (alternatively, this directly follows from Cauchy-Schwarz). Then $2(n+19) \geq n^2 \iff n^2-2n-38 \leq 0$. The largest positive integer n satisfying this inequality is 7, so $n \leq 7$. It can be checked that the equations $x+y = 7$ and $x^2+y^2 = 26$ have a real solution (x, y), so the answer is $\boxed{7}$. □

22. Our configuration should resemble the diagram below (the actual labeling of the points and lines is not important). We will label segments of equal length with x, y, and z:

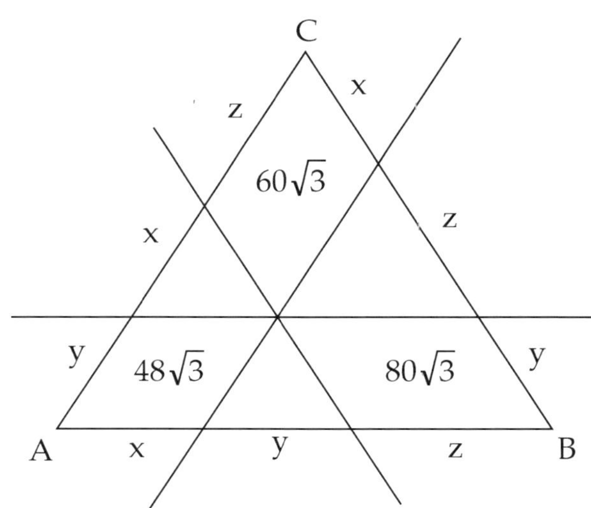

Because the three parallelograms each contain 60° and 120° angles, we can express the areas of the parallelograms easily in terms of x, y, and z:

$$xy \sin 60^\circ = 48\sqrt{3} \implies xy = 96$$
$$xz \sin 60^\circ = 60\sqrt{3} \implies xz = 120$$
$$yz \sin 60^\circ = 80\sqrt{3} \implies yz = 160$$

We can make the observation that $48 = 6 \times 8$, $60 = 6 \times 10$, and $80 = 8 \times 10$ to find x, y, z. In this case, we have $x = 6\sqrt{2}$, $y = 8\sqrt{2}$, and $z = 10\sqrt{2}$, so $AB = 24\sqrt{2}$. The area of equilateral $\triangle ABC$ is $\frac{(24\sqrt{2})^2\sqrt{3}}{4} = \boxed{288\sqrt{3}}$. □

23. Let $f(n) = \lfloor \frac{n}{3} \rfloor + \lfloor \frac{n}{4} \rfloor + \lfloor \frac{n}{5} \rfloor$. We first note that not every positive integer can be written in the form $f(n)$ for some positive integer n (e.g., if $n = 60$, then $f(60) = 20 + 15 + 12 = 47$, but if $n = 59$, then $f(59) = 19 + 14 + 9 = 44$, so 45 and 46 cannot be expressed in this way).

The above observation clues us to observe that if n is a multiple of 60, then two numbers are skipped, and if n is a multiple of 12, 15, or 20 but not 60, then one number is skipped (e.g., $f(12) = 9$ and $f(11) = 7$). Since $f(1278) = 1000$, we can find the multiples of 12, 15, 20, and 60 less than 1278.

There are $\lfloor \frac{1278}{12} \rfloor = 106$ multiples of 12, $\lfloor \frac{1278}{15} \rfloor = 85$ multiples of 15, and $\lfloor \frac{1278}{20} \rfloor = 63$ multiples of 20 less than 1278. There are $\lfloor \frac{1278}{60} \rfloor = 21$ multiples of 60 less than 1278.

Adding, we obtain $106 + 85 + 63 = 254$ multiples of 12, 15, and 20, or 254 numbers out of the first 1000 which are skipped. However, we overcounted the multiples of 60; each multiple of 60 is counted exactly three times, but we want to count each multiple of 60 twice (as each multiple of 60 skips two numbers). The number of numbers which are skipped is $254 - 21 = 233$. Hence the number of integers expressible in the form $\lfloor \frac{n}{3} \rfloor + \lfloor \frac{n}{4} \rfloor + \lfloor \frac{n}{5} \rfloor$ for some positive integer n is $1000 - 233 = \boxed{767}$. □

24. The given circle is centered at $(2, 3)$ and has radius 1:

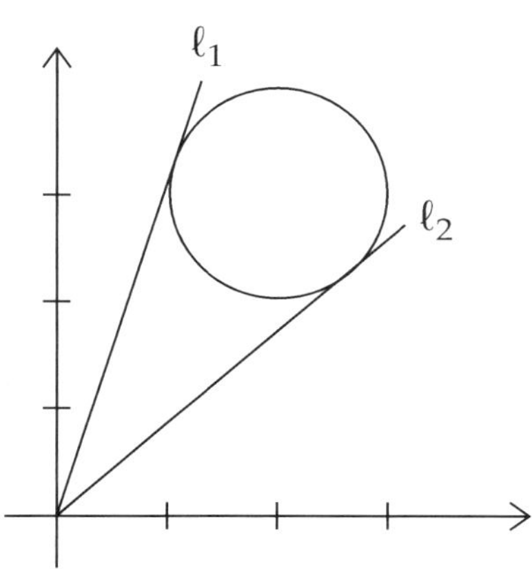

Consider the line $y = mx$ for some m, or $mx - y = 0$. The distance from ℓ_1 or ℓ_2 to the point $(2, 3)$ is 1. Using the formula for the distance from a line to a point, we have

$$\frac{|2m - 3 + 0|}{\sqrt{m^2 + (-1)^2}} = 1$$
$$|2m - 3| = \sqrt{m^2 + 1}$$

Square both sides to obtain $4m^2 - 12m + 9 = m^2 + 1$, or $3m^2 - 12m + 8 = 0$. This is quadratic in m, and has two real roots, which are the slopes of ℓ_1 and ℓ_2. By Vieta's formulas, the product of the slopes is $\boxed{\frac{8}{3}}$. □

25. Given positive integer n, let $f(n)$ denote the minimum possible length of such a sequence satisfying the given properties. For example, $f(28) = 7$ since there exists a length 7 sequence but no length 6 or shorter sequence. The answer is the number of integers n such that $f(n) \le 12$. We compute $f(n)$ for small n:

n	$f(n)$	Sequence
1	1	1
2	2	1, 2
3	3	1, 2, 3
4	3	1, 2, 4
5	4	1, 2, 4, 5
6	4	1, 2, 3, 6
7	5	1, 2, 3, 6, 7
8	4	1, 2, 4, 8
9	5	1, 2, 4, 8, 9
10	5	1, 2, 4, 5, 10
11	6	1, 2, 4, 5, 10, 11
12	5	1, 2, 3, 6, 12

The above table suggests the following observations: for odd n, $f(n) = f(n-1) + 1$, and for even n, $f(n) = f\left(\frac{n}{2}\right) + 1$. We formalize this as follows.

Claim: Let $d(n)$ denote the number of digits of n when written in base 2, and let $s(n)$ denote the number of 1's when n is written in base 2. Then $f(n) = d(n) + s(n) - 1$.

Proof. We prove by strong induction on n. We can check that the base cases ($n = 1$, $n = 2$, etc.) hold. Suppose $f(k) = d(k) + s(k) - 1$ for all $k < n$, where $n \ge 2$. We wish to show $f(n) = d(n) + s(n) - 1$.

If n is odd, then any sequence ending in n must end with $n - 1$, then n, in which we obtain $f(n) = f(n-1) + 1$. By induction hypothesis, $f(n-1) = d(n-1) + s(n-1) - 1$, so $f(n) = d(n-1) + s(n-1)$. As $n - 1$ is even, $d(n-1) = d(n)$, and $s(n-1) = s(n) - 1$. Substituting, we obtain $f(n) = d(n) + s(n) - 1$, proving this case.

If n is even, then any sequence ending in n must either end in $\dots, \frac{n}{2}, n$, or $\dots, n-2, n-1, n$. Hence we establish that $f(n) = \min(f\left(\frac{n}{2}\right) + 1, f(n-2) + 2)$. By induction hypothesis, and

using the fact that n is even:

$$\begin{aligned} f(n) &= \min\left(d\left(\frac{n}{2}\right)+s\left(\frac{n}{2}\right), d(n-2)+s(n-2)+1\right) \\ &= \min(d(n)+s(n)-1, d(n-2)+s(n-2)+1) \end{aligned}$$

since $d\left(\frac{n}{2}\right) = d(n)-1$ and $s\left(\frac{n}{2}\right) = s(n)$. This immediately implies $f(n) \leq d(n)+s(n)-1$. Suppose $f(n) < d(n)+s(n)-1$. Further, note by induction hypothesis that $f(n-2) = d(n-2)+s(n-2)-1$. Since $f(n) \leq f(n-2)+2$ (as a candidate solution is to construct the minimum sequence ending in $n-2$, then add 1 twice), this implies $f(n) \leq d(n-2)+s(n-2)+1$, or in other words, $d(n-2)+s(n-2)+1$ cannot be less than $f(n)$. Hence $f(n) = d(n)+s(n)-1$ as desired. □

Now the problem reduces to finding the number of positive integers n such that $2 \leq d(n)+s(n)-1 \leq 12$, or equivalently, $3 \leq d(n)+s(n) \leq 13$. Here, observe that in the binary representation of n, each 0 contributes exactly 1 to the sum $d(n)+s(n)$, and each 1 contributes exactly 2 to the sum $d(n)+s(n)$. Hence, we can put the set of good integers in bijection (one-to-one correspondence) with sequences of 1's and 2's whose sum is between 1 and 11 inclusive as follows: given a good integer n, convert it to binary, add 1 to every digit, then drop the leading 2.

The number of sequences of 1's and/or 2's whose sum is k is equal to F_{k+1} where F_k is the kth Fibonacci number (a classic exercise in recursion). Sum from $k=1$ to $k=11$ to obtain the answer:

$$\begin{aligned} \sum_{k=1}^{11} F_{k+1} &= F_2+F_3+\ldots+F_{12} \\ &= (F_1+F_2+\ldots+F_{12})-1 \\ &= (F_{14}-1)-1 \\ &= F_{14}-2 \\ &= \boxed{375} \end{aligned}$$

□

Solutions for Practice Exam 4.

1. We have $x + y = 24$, $x + z = 12$, and $y + z = 8$. Adding all three equations and dividing by 2 gives us $x + y + z = 22$. Since $x + y = 24$, we obtain $z = \boxed{-2}$. □

2. The smallest prime factor of such a number n must be at least 11. We can list all *almost prime* numbers in order of its smallest prime factor: 11^2, 11×13, 11×17, 11×19, 11×23, 13×13, 13×17, 13×19, 13×23, 17×17, or $\boxed{10}$ numbers. □

3. No matter what the first card drawn is, there are 12 out of 51 cards which are of the same suit as the first card. The probability is $\frac{12}{51} = \boxed{\frac{4}{17}}$. □

4. Note that we only need to compare the area of one square with the area of one octagon. This is more easily seen in the following picture (by considering the dashed square):

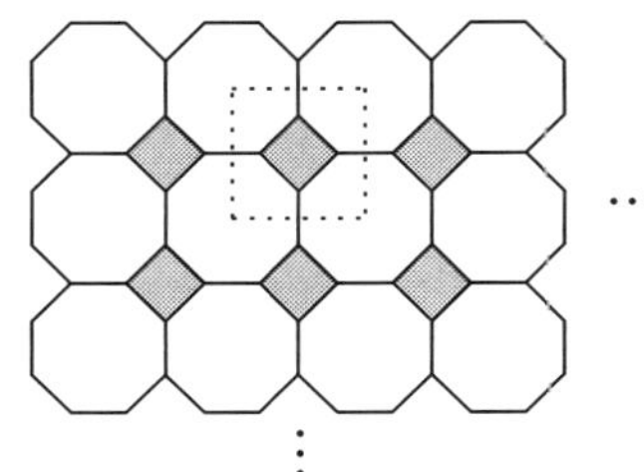

Without loss of generality, suppose the side length of the square (and the octagon) is 1. The area of the square is 1, and the area of the octagon is $2 + 2\sqrt{2} \approx 4.82$.

We want the percentage of the total area covered by the squares. This is approximately $\frac{1}{1+(2+2\sqrt{2})} \approx \frac{1}{5.82} \approx \boxed{17\%}$ (careful: this is *not* the ratio of the square's area to the octagon's area, but rather the ratio of the square's area to the sum of these two areas). □

5. We observe that $(a, b, c) = (k^2 - 1, k^2, k(k^2 - 1)!)$ is a solution for all positive integers k. Therefore there are $\boxed{\text{Infinitely many}}$ solutions. □

6. Since $20^{20} = 2^{40}5^{20}$, the number of positive divisors of 20^{20} is $(40 + 1)(20 + 1) = 861$.

   We use complementary counting, by finding the number of divisors of 20^{20} which are divisible by 20. Note that any divisor of 20^{20} is of the form 2^a5^b where a and b are integers, $0 \le a \le 40$, and $0 \le b \le 20$. Such a divisor is divisible by 20 if and only if $a \ge 2$ and $b \ge 1$. Then there are $39 \times 20 = 780$ divisors of 20^{20} which are divisible by 20.

   Since we are looking for the number of divisors *not* divisible by 20, the answer is $861 - 780 = \boxed{81}$. □

7. Observe that such a 4-element subset $T \subset S$ is of the form $T = \{a, a + d, a + 2d, a + 3d\}$, in which the smallest and largest elements differ by a multiple of 3. We claim that we can choose the largest and smallest elements of T (which must be congruent modulo 3), which will uniquely determine T.

   There are 10 elements of S which are congruent to $0 \pmod 3$, 10 elements congruent to $1 \pmod 3$, and 10 elements congruent to $2 \pmod 3$. For all $i \in \{0, 1, 2\}$, if the smallest and largest elements of T are congruent to $i \pmod 3$, then there are $\binom{10}{2} = 45$ ways to choose the smallest and largest elements of T, which uniquely determines the subset T. Hence there are $3 \times 45 = \boxed{135}$ four-element subsets T satisfying this property. □

8. Let r denote the radius of the smaller sphere. The configuration will look something like this:

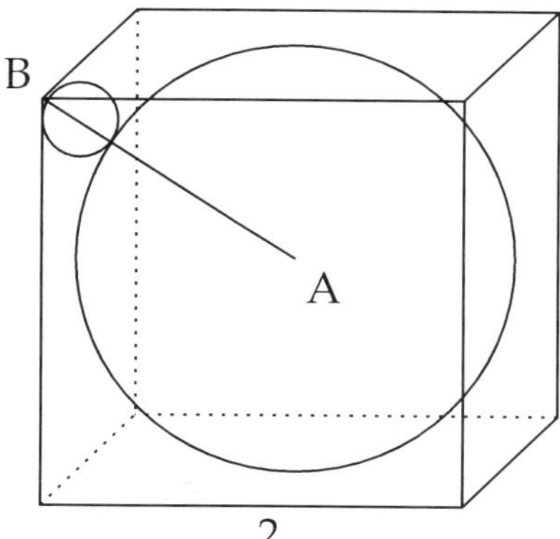

The length of segment AB is half the space diagonal, or $\sqrt{3}$. Further, AB passes through the center of the smaller sphere. Letting O be the center of the smaller sphere and P be the point of tangency of the smaller sphere with the larger sphere, we see that $AP = 1, PO = r$, and $OB = r\sqrt{3}$, so $AB = \sqrt{3} = 1 + r + r\sqrt{3}$. Solving for r, we get $r = \frac{\sqrt{3}-1}{\sqrt{3}+1} = \boxed{2 - \sqrt{3}}$. □

9. Recall the quadratic formula: if $ax^2 + bx + c = 0$, then $x = \frac{-b \pm \sqrt{b^2-4ac}}{2a}$. If both solutions are real, then the positive difference between these solutions is $\frac{\sqrt{b^2-4ac}}{a}$. Substituting $a = 1$, $b = 8$, and $c = -84$, we obtain $\sqrt{8^2 + 4(84)} = \boxed{20}$. □

10. We can compute $5^{100} \pmod 8$ and $\pmod 9$; by the Chinese remainder theorem, this gives us enough information to determine $5^{100} \pmod{72}$.

    Note that $5^2 = 25 \equiv 1 \pmod 8$, so $5^{100} \equiv (5^2)^{50} \equiv 1 \pmod 8$.

    Also, $5^3 = 125 \equiv -1 \pmod 9$, so $5^{99} \equiv (5^3)^{33} \equiv (-1)^{33} \equiv -1 \pmod 9$. Multiplying by 5, we have $5^{100} \equiv -5 \equiv 4 \pmod 9$.

    Hence the number 5^{100} is congruent to $1 \pmod 8$ and $4 \pmod 9$, so we can determine $5^{100} \pmod{72}$. The only number in $[0, 71]$ congruent to $1 \pmod 8$ and $4 \pmod 9$ is $\boxed{49}$.

    Alternate solution: We have $\varphi(72) = 72 \times \frac{1}{2} \times \frac{2}{3} = 24$, where $\varphi(\cdot)$ is Euler's totient function. By Euler's totient theorem, we have $5^{24} \equiv 1 \pmod{72}$. Then $5^{96} \equiv 1 \pmod{72} \implies 5^{100} \equiv 5^4 \equiv \boxed{49} \pmod{72}$. □

11. We can find the number of ways to form a monetary amount which is a multiple of 5 cents (less than 47 cents), using nickels, dimes, and/or quarters. This follows as any combination of nickels, dimes, and/or quarters will uniquely determine the number of pennies to use.

    If the cashier uses 1 quarter, then he can use zero, one or two dimes. We obtain the following combinations: QDD, QD, QDN, QDNN, Q, QN, QNN, QNNN, QNNNN.

    If the cashier uses 0 quarters, then he can use 0, 1, 2, 3, or 4 dimes. We obtain the following combinations: DDDD, DDDDN, DDD, DDDN, DDDNN, DDDNNN, DD, DDN, ..., DDNNNNN, D, DN, ..., DNNNNNNN, (no dimes or nickels), N, ..., NNNNNNNNN, giving $2 + 4 + 6 + 8 + 10 = 30$ combinations.

Altogether, we get $9 + 30 = \boxed{39}$ combinations. □

12. Rewrite the given equation as $\frac{\sin^2\alpha+\cos^2\alpha}{\sin\alpha\cos\alpha} = \frac{13}{6}$. Since $\sin^2\alpha + \cos^2\alpha = 1$, this implies $\sin\alpha\cos\alpha = \frac{6}{13}$, or $\sin 2\alpha = \frac{12}{13}$ (using the identity $\sin 2\alpha = 2\sin\alpha\cos\alpha$).

Consider the following 5-12-13 triangle. Note that $\alpha \in [0, \frac{\pi}{4}]$, so $2\alpha \in [0, \frac{\pi}{2}]$.

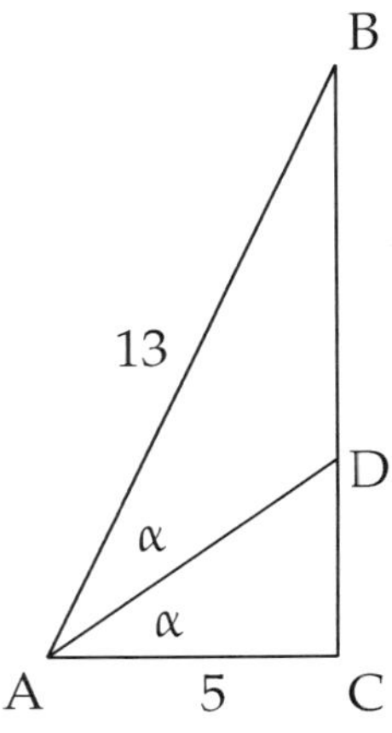

By the angle bisector theorem, $CD = \frac{5}{18} \cdot 12 = \frac{10}{3}$, so by the Pythagorean theorem, $AD = \frac{5\sqrt{13}}{3}$. Then $\sin\alpha = \frac{CD}{AD} = \frac{10}{5\sqrt{13}} = \boxed{\frac{2\sqrt{13}}{13}}$. □

13. By the change of base formula, $\log_{225} 256 = \frac{\log_2 256}{\log_2 225} = \frac{8}{\log_2 225}$. We can write $\log_2 225$ as $\log_2(15^2) = 2\log_2 15 = 2(\log_2 3 + \log_2 5) = 2(a+b)$. Substituting, the desired expression equals $\frac{8}{2(a+b)} = \boxed{\frac{4}{a+b}}$. □

14. We use the Chinese remainder theorem. Consider the system of congruences

$$\begin{aligned} n &\equiv r_1 \pmod{9} \\ n &\equiv r_2 \pmod{10} \end{aligned}$$

where $0 \leq r_1 < 9$ and $0 \leq r_2 < 10$. By the Chinese remainder theorem, we can uniquely determine $n \pmod{90}$, for a fixed choice of r_1, r_2.

Here, we want r_1 and r_2 to be different. There are 9 ways to choose r_1; given this, there are $10 - 1 = 9$ ways to choose r_2. This gives $9 \times 9 = 81$ possible residues, out of 90. Hence, for every 90 consecutive numbers, exactly 81 of them leave different remainders. Since there are 900 3-digit numbers, the desired answer is $\boxed{810}$. □

15. The main idea behind the solution is that the expected value of any one die roll is $\frac{7}{2}$; the expected value of n (the number of dice rolls added in Richard's sum) is also $\frac{7}{2}$, so the answer is $\frac{7}{2} \times \frac{7}{2} = \boxed{\frac{49}{4}}$.

Here, we present a more formal treatment of the problem. Let $X_1, X_2, \ldots, X_6$ be random variables such that $X_i = 1$ if $n \geq i$, and $X_i = 0$ otherwise. In particular, $X_1 = 1$ with probability 1, and $X_6 = 1$ with probability $\frac{1}{6}$. More generally, $\mathbb{E}(X_i) = \frac{7-i}{6}$.

Consider the following equivalent scenario: instead of rolling n fair six-sided dice, Richard rolls 6 fair six-sided dice and only adds the first n dice rolls. Let $Y_1, Y_2, \ldots, Y_6$ denote the values of these dice rolls. We see that Y_1 is always included in the sum, Y_2 is included in the sum with probability $\frac{5}{6}$, ..., Y_6 is included in the sum with probability $\frac{1}{6}$. Our desired answer is

$$\mathbb{E}(X_1Y_1 + X_2Y_2 + \ldots + X_6Y_6).$$

By linearity of expectation, the above expression equals

$$\sum_{i=1}^{6} \mathbb{E}(X_iY_i) = \sum_{i=1}^{6} \mathbb{E}(X_i)\mathbb{E}(Y_i)$$

using the fact that X_i and Y_i are independent. Since $\mathbb{E}(Y_i) = \frac{7}{2}$, and $\mathbb{E}(X_i) = \frac{7-i}{6}$, we obtain

$$\frac{7}{2}\left(1 + \frac{5}{6} + \frac{4}{6} + \ldots + \frac{1}{6}\right) = \boxed{\frac{49}{4}}.$$

□

16. Let $OA = a$ and $AB = b$, so that $2a+2b = 16 \iff a+b = 8$. Observe that since $OABC$ is a rectangle, $AC = OB = 6$. The perimeter of the shaded region is $3\pi+6+(6-a)+(6-b) = 3\pi+18-(a+b)$. Since $a+b = 8$, the perimeter is $3\pi+18-8 = 3\pi+10$, so $m = \boxed{10}$. □

17. Substitute $x = \sin\alpha$ where $\alpha \in [0, \frac{\pi}{2}]$. Then $1 - x^2 = 1 - \sin^2\alpha = \cos^2\alpha$, and since $\cos\alpha \geq 0$, we have $\sqrt{1-x^2} = \cos\alpha$. Hence we are maximizing $\sin\alpha\cos\alpha = \frac{1}{2}\sin 2\alpha$ on the interval $\alpha \in [0, \frac{\pi}{2}]$. The maximum of $\frac{1}{2}\sin 2\alpha$ is $\boxed{\frac{1}{2}}$, attained when $\alpha = \frac{\pi}{4}$. □

18. Any divisor of 10^{10} is of the form 2^a5^b, where $0 \leq a \leq 10$ and $0 \leq b \leq 10$. Since $2, 5 \equiv 2 \pmod 3$, we see that $2^a5^b \equiv 1 \pmod 3$ if and only if $a+b$ is even.

    Hence we can reduce the problem to finding the number of integer pairs (a, b) such that $0 \leq a, b \leq 10$ and $a+b$ is even. If a and b are both odd, there are $5 \times 5 = 25$ pairs (a, b). If a and b are both even, there are $6 \times 6 = 36$ pairs. Adding, we get $25 + 36 = \boxed{61}$ pairs, or divisors of 10^{10} congruent to $1 \pmod 3$. □

19. We have

$$\begin{aligned} y^2 - x^2 = (y+x)(y-x) &= 111 + 6y \\ \implies y^2 - x^2 - 6y + 9 &= 120 \\ \implies (y-3)^2 - x^2 &= 120 \\ \implies (y-3+x)(y-3-x) &= 120 \end{aligned}$$

Since 120 and $(y-3+x) - (y-3-x) = 2x$ are even, both $(y-3+x)$ and $(y-3-x)$ should be even. Say $y-3-x = 2k$, then $y-3+x = 2k+2x$. By simplifying,

$$k(k+x) = 30$$

For each divisor (positive or negative) of 30 for k, we get a integer pair (x, y), thus the number of pairs is $2 \times 8 = \boxed{16}$. □

20. Let $BD = x$, so that $AD = 2x$ and $DC = 3x$. Then $\triangle ABD$ has side lengths x, $2x$, and 2, and $\triangle ADC$ has side lengths $2x$, $3x$, and 4.

Using the area formula $[\triangle ABC] = \frac{abc}{4R}$, we can find the ratio of R_1 and R_2. First, note that $[\triangle ADC] = 3[\triangle ABD]$, so let $[\triangle ABD] = K$ and $[\triangle ADC] = 3K$. Then we have

$$K = \frac{x(2x)(2)}{4R_1} = \frac{x^2}{R_1}$$
$$3K = \frac{(2x)(3x)(4)}{4R_2} = \frac{6x^2}{R_2}$$

Cross-multiplying yields $R_1 = \frac{x^2}{K}$ and $R_2 = \frac{2x^2}{K}$ so $\frac{R_1}{R_2} = \boxed{\frac{1}{2}}$.

Alternate solution: Let $\angle ADB = \alpha$, so that $\angle ADC = \pi - \alpha$. Using the extended law of sines on $\triangle ABD$ and $\triangle ADC$:

$$\frac{2}{\sin \alpha} = 2R_1 \qquad \frac{4}{\sin(\pi - \alpha)} = 2R_2$$

However $\sin \alpha = \sin(\pi - \alpha)$, so we immediately obtain $R_2 = 2R_1$, or $\frac{R_1}{R_2} = \boxed{\frac{1}{2}}$. □

21. Instead of a, b, c, we can use x, y, z so that the geometric interpretation is clearer. The graph of $x + 2y + 3z = 1$ is a plane, and $x^2 + y^2 + z^2$ represents the square of the distance from the point (x, y, z) to the point $(0, 0, 0)$. Using the formula for the distance from a point to a plane, we have

$$D = \frac{|0 + 0 + 0 - 1|}{\sqrt{1^2 + 2^2 + 3^2}} = \frac{1}{\sqrt{14}}$$

Since we want the distance squared, our answer is $D^2 = \boxed{\frac{1}{14}}$.

Alternate solution: By the Cauchy-Schwarz inequality, we have

$$(a^2 + b^2 + c^2)(1^2 + 2^2 + 3^2) \geq (a + 2b + 3c)^2 = 1$$

This implies $a^2 + b^2 + c^2 \geq \frac{1}{1^2+2^2+3^2} = \frac{1}{14}$. Equality occurs when (a, b, c) is a scalar multiple of $(1, 2, 3)$; since $a + 2b + 3c = 1$, equality occurs when $(a, b, c) = \left(\frac{1}{14}, \frac{2}{14}, \frac{3}{14}\right)$. □

22. Consider triangle $\angle ABC$. By the inscribed angle theorem, since $\overline{AB}$ is a diameter, we have that $\angle ACB = 90°$, so $AC = \sqrt{8^2 - 2^2} = 2\sqrt{15}$.

Also, observe that $\triangle BCD$ is isosceles with $\angle DBC = \angle CBD = \alpha$; this implies $\angle CAD = \angle CAB = \alpha$ as well. We know that $\cos \alpha = \frac{AC}{AB} = \frac{\sqrt{15}}{4}$. We will apply the law of cosines on $\triangle ACD$ to find AD:

$$DC^2 = AD^2 + AC^2 - 2 \cdot AD \cdot AC \cos \alpha$$
$$4 = AD^2 + 60 - 2 \cdot AD \cdot 2\sqrt{15} \cdot \frac{\sqrt{15}}{4}$$
$$AD^2 - 15AD + 56 = 0$$
$$(AD - 7)(AD - 8) = 0$$

Since AD is a chord of the circle which is not the diameter (as shown in the given figure), $AD < 8$. Hence the length of $\overline{AD}$ is $\boxed{7}$. □

23. We list out the first several terms of the sequence. As clued by the question, we also list their corresponding base-3 representations.

1	1
3	10
9	100
10	101
27	1000
28	1001
30	1010
81	10000
82	10001
83	10010
90	10100
91	10101
243	100000

We make the following observation: for each $k \geq 0$, the number of terms in the sequence whose base-3 representation contains k digits is exactly F_k, where F_n is the nth Fibonacci number, defined by $F_0 = 0$, $F_1 = 1$, and $F_n = F_{n-1} + F_{n-2}$ for $n \geq 2$ (try to prove this!).

Since $F_1 + F_2 + F_3 + ... + F_9 = F_{11} - 1 = 88$, the 100th term will have 10 digits in its base-3 representation, and the 89th term is $1000000000_3 = 3^9 = 19683$. It follows that the 100th term is $1000010100_3 = 19683 + 81 + 9 = \boxed{19773}$.

Alternate solution: We use a unique property about the Fibonacci numbers: every positive integer can uniquely be written as a sum of one or more nonconsecutive terms from the set $\{1, 2, 3, 5, 8, ...\}$ (this is known as *Zeckendorf's theorem*, which can be used to generate a *Zeckendorf representation* of a positive integer). The base-3 representations of the terms of this sequence correspond exactly to the Zeckendorf representations of 1, 2, 3, 4, ...The Zeckendorf representation of 100 is $89 + 8 + 3$ which corresponds to 1000010100_3, from which the answer follows. □

24. We solve this problem using recursion. Let a_n denote the number of sequences whose first term is 1, last term is n, with positive difference between any two consecutive terms at least 3. Then $a_1 = 1$, $a_2 = a_3 = 0$. To find a_n, we can construct such a sequence by appending n to any valid sequence ending in a number less than or equal to $n - 3$. Hence

$a_n = a_{n-3} + a_{n-4} + a_{n-5} + ... + a_1$ for $n \geq 4$. We may compute $a_4 = a_5 = a_6 = 1$, $a_7 = 2$, and so on.

We can simplify the recursion by noticing that $a_{n-4} + a_{n-5} + ... + a_1 = a_{n-1}$, for $n \geq 5$. Hence we have the recurrence $a_n = a_{n-1} + a_{n-3}$ for $n \geq 5$. We can compute values up to a_{20} to obtain $a_{20} = \boxed{277}$. □

25. Rewrite the desired expression as $\frac{r^3t+s^3r+t^3s}{rst}$. By Vieta's formulas, $rst = (-1)^3(1) = -1$, so it suffices to find $r^3t + s^3r + t^3s$.

We can substitute $r^3 = 5r - 1$, $s^3 = 5s - 1$, $t^3 = 5t - 1$ as r, s, and t are the roots. This simplifies to finding $(5r-1)t+(5s-1)r+(5t-1)s = 5(rt+rs+st)-(r+s+t)$. Applying Vieta's formulas again, we have $rt+rs+st = -5$ and $r+s+t = 0$, so $r^3t+s^3r+t^3s = 5(-5) = -25$.

The answer is therefore $\frac{-25}{-1} = \boxed{25}$. □

1. If he buys four bags, the price paid is $\$2.50 \times 4 = \$10$. If five bags are bought, the price paid is $0.7 \times \$2.50 \times 5 = \$8.75$. He saves $\$10 - \$8.75 = \boxed{\$1.25}$ if he buys five bags instead of four. □

2. We will assume a and b are positive for now. If (a, b) is a solution where $a, b > 0$, then $(-a, b)$, $(a, -b)$, and $(-a, -b)$ are also solutions.

   Factor the left-hand side using difference of squares to get $(a+b)(a-b) = 2020$. The prime factorization of 2020 is $2^2 \times 5 \times 101$, and we see that $a+b$ and $a-b$ must be two positive factors (under the assumption $a, b > 0$) with the same parity. Hence $a+b$ and $a-b$ must both be even, and we observe that $(a+b, a-b) = (2, 1010)$, $(10, 202)$, $(202, 10)$, or $(1010, 2)$. The first two cases do not work since $a+b$ is necessarily greater than $a-b$. Hence $(a+b, a-b) = (202, 10)$ or $(1010, 2)$ giving solutions $(a, b) = (106, 96)$ and $(506, 504)$.

   However, since a and b are not necessarily positive, we see that the solutions are $(a, b) = (\pm 106, \pm 96)$ and $(\pm 506, \pm 504)$, giving a total of $2 \times 4 = \boxed{8}$ ordered pairs. □

3. Two squares drawn in the same plane intersect at most 8 points. There are $\binom{4}{2} = 6$ ways to choose a subset of two squares, so the number of points of intersection is at most $8 \times 6 = 48$. Indeed, $\boxed{48}$ points is possible, as shown in the following picture:

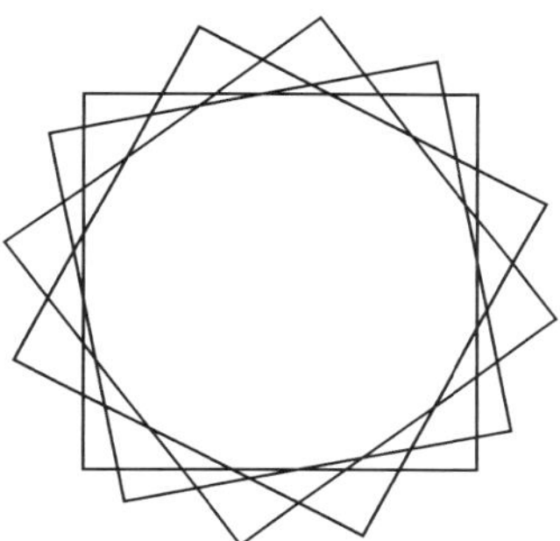

□

4. We see that $\triangle AGH$ is isosceles with $AH = HG$, and $m\angle H = 135°$ as it is one of the interior angles of the octagon. Then $m\angle AGH = \frac{180° - 135°}{2} = \boxed{22.5°}$. □

5. We have a few cases: either i) $x^2 - 3x + 1 = 1$, or ii) $x^2 - 3x + 1 = -1$ and $x + 1$ is even, or iii) $x + 1 = 0$ and $x^2 - 3x + 1 \neq 0$.

   Case i) gives $x^2 - 3x = 0$, or $x = 0, 3$. Case ii) gives $x^2 - 3x + 2 = 0 \implies x = 1, 2$; however $x = 2$ does not work as the exponent $x + 1$ is odd. So case ii) gives one solution $x = 1$. Case iii) gives $x = -1$, which works.

   Hence the solution set is $x \in \{-1, 0, 1, 3\}$, so there are $\boxed{4}$ integer solutions. □

6. We can immediately rule out choices B and D by considering $n = 10$, which is deficient. The numbers 20 and 100 are not deficient as $\sigma(20) - 20 = 1 + 2 + 4 + 5 + 10 > 20$ and $\sigma(100) - 100 = 117 > 100$.

   Choice C is incorrect as well; consider $n = 5$ which is deficient (all primes are deficient). Then $5^2 - 1 = 24$ which is not deficient.

Choice A is incorrect as well; consider $n = 13$, which is deficient. Then $\lfloor \frac{n}{2} \rfloor = 6$ which is not deficient (6 is a *perfect number*). However if n is even and deficient, then $\frac{n}{2}$ is also deficient.

We have ruled out choices A–D, so the answer is $\boxed{\text{None of the preceding}}$. □

7. There are 15 representatives in total. So there are $\binom{15}{2}$ handshakes without any restriction. However we want each representative to shake hands with all the representatives from companies other than their own, so we need to exclude $\binom{3}{2}$ handshakes per company. Thus our answer is $\binom{15}{2} - 5 \cdot \binom{3}{2} = \boxed{90}$.

   Alternate solution: Each person shakes hands with exactly 12 people, so $15 \times 12 = 180$ "handshakes" occur. However, we counted each handshake between any two people twice, so we divide by 2, giving $\frac{180}{2} = \boxed{90}$ handshakes. □

8. The slope of $\overline{YX}$ is $\frac{1-2}{3-0} = -\frac{1}{3}$; since $\overline{YX} \perp \overline{YZ}$, the slope of $\overline{YZ}$ must be 3. Then $z = 2 + 3(2) = \boxed{8}$. □

9. Suppose the first term is a. Then the second term is $a+1$, and the third term is $(a+1)+4 = a+5$. Since the terms are in geometric progression, we have $(a+1)^2 = a(a+5) \implies a^2 + 2a + 1 = a^2 + 5a \implies a = \frac{1}{3}$. Then the sequence is $\frac{1}{3}, \frac{4}{3}, \frac{16}{3}, \frac{64}{3}, \ldots$, and the fourth term is $\boxed{\dfrac{64}{3}}$. □

10. Assume n is the last page of the book. Then we have

$$1 + \ldots + n = \frac{n \cdot (n+1)}{2} < 2020 \text{ and } 1 + \ldots + (n+1) = \frac{(n+1) \cdot (n+2)}{2} > 2020.$$

    Since n is an integer, we have $n \le 63$ in the first inequality, and $n + 1 \ge 64$ in the second inequality. Thus $n = \boxed{63}$. □

11. The possible remainders (residues) upon division by 3 are 0, 1, 2. Given a 3-element subset $S \subset \{1, 2, 3, \ldots, 12\}$, there are only two possibilities: all elements in S are the same modulo 3, or all elements in S are different modulo 3.

    If all elements in S are the same modulo 3, this gives $3 \times \binom{4}{3} = 12$ possible subsets. If all elements in S are different modulo 3, this gives $4 \times 4 \times 4 = 64$ subsets. Altogether we obtain $12 + 64 = \boxed{76}$ subsets. □

12. Note that A, B, C, and D are on the line segment, in that order. We have $AB = 3BD$ and $AC = 7CD$.

    Without loss of generality, suppose $AD = 8$. Then $AB = 6$, $BD = 2$, $AC = 7$, and $CD = 1$, so $BC = 7 - 6 = 1$, and $\frac{AD}{BC} = \boxed{8}$. □

13. Let $a = 5^x$ and $b = x^2$. The equation can be written as $ab + 125 = 25a + 5b \iff ab - 25a - 5b + 125 = 0$. We can factor the left-hand side to get $(a-5)(b-25) = 0$. Then either $a = 5$ or $b = 25$, giving the real solutions $x = 1$ or $x = \pm 5$. Hence there are $\boxed{3}$ real solutions. □

14. We claim that the answer is simply $\frac{1}{1000}+\frac{2}{1000}+\frac{3}{1000}+\cdots+\frac{999}{1000}$. Every fraction of the form $\frac{n}{1000}$ can be reduced to a fraction of the form $\frac{a}{b}$ where b is a divisor of 1000. Moreover, no other fractions appear in this set.

    Hence the sum of the elements of S is $\frac{1}{1000}+\frac{2}{1000}+\frac{3}{1000}+\cdots+\frac{999}{1000}=\frac{1}{1000}(1+2+3+\cdots+999)=\frac{1}{1000}\times\frac{999\times1000}{2}=\boxed{\frac{999}{2}}$. □

15. Consider the three-digit number $\overline{ABC}$ modulo 7, where $1\le A,B,C\le 6$. We make the following observations:

    - If the two-digit number $\overline{AB}$ is divisible by 7, then there is no value of C for which the three-digit number $\overline{ABC}$ is divisible by 7.
    - If the two-digit number $\overline{AB}$ is not divisible by 7, then there is exactly one value of C for which the three-digit number $\overline{ABC}$ is divisible by 7.

    The second observation can be shown by noting that if $\overline{AB}\equiv k \pmod 7$ where $k\in\{1,2,\ldots,6\}$, then $\overline{AB0}\equiv 3k \pmod 7$, so set $C=(-3k) \pmod 7$.

    Hence we can reduce the problem to finding the number of two-digit numbers of the form $\overline{AB}$ where $1\le A,B\le 6$ which are not divisible by 7. This equals 36 minus the number of numbers $\overline{AB}$ which are divisible by 7. This can be counted manually: 14, 21, 35, 42, 56, 63. Then $36-6=30$ two-digit numbers of the form $\overline{AB}$ are not divisible by 7, which implies that 30 three-digit numbers of the form $\overline{ABC}$ are divisible by 7. There are $6^3=216$ possible outcomes, so the probability is $\frac{30}{216}=\boxed{\frac{5}{36}}$. □

16. Let $A_1A_3=s$. Using the law of cosines on $\triangle A_1A_2A_3$, we have $s^2=1^2+1^2-2\cos 150^\circ=1+1-2\left(-\frac{\sqrt{3}}{2}\right)=2+\sqrt{3}$.

    The area of a regular hexagon with side length s is $6\times\frac{s^2\sqrt{3}}{4}$. In this case, $s^2=2+\sqrt{3}$, so the area of regular hexagon $A_1A_3A_5A_7A_9A_{11}$ is $6\times\frac{(2+\sqrt{3})\sqrt{3}}{4}=\boxed{\frac{9+6\sqrt{3}}{2}}$. □

17. Since $(x-y)^2\ge 0$, and so $x^2+y^2\ge 2xy$, for all real numbers x and y, then

$$3=x^2+xy+y^2\ge 3xy.$$

    Thus $1\le xy$ and so $x^2+y^2=3-xy\ge 2$. Hence x^2+y^2 cannot be $\boxed{1}$. □

18. Note that $17^{128}-1$ can be factored as

$$17^{128}-1=\left(17^{64}+1\right)\left(17^{32}+1\right)\left(17^{16}+1\right)\left(17^{8}+1\right)\left(17^{4}+1\right)\left(17^{2}+1\right)(17+1)(17-1).$$

    All factors except $(17-1)$ are divisible by 2 but not 4, that is, they have only one power of 2. On the other hand $(17-1)=16=2^4$. So the maximum value of n is $\boxed{11}$. □

19. Notice that, if the base-8 representation of n contains the digit "0", then the base-4 representation of n will also contain the digit 0. This is because we can consider n written in binary, in which the base-4 and base-8 representations can be easily formed by taking digits two or three at a time.

    Hence, all we need is the number of positive integers less than or equal to 1,000 whose base-8 representation contains the digit "0." Note that $1000 = 1750_8$. We can use complementary counting, by counting the number of integers whose base 8 representation does not contain the digit 0. There are 7 one-digit (base 8) numbers, $7^2 = 49$ two-digit numbers, $7^3 = 343$ three-digit numbers. To find the number of four-digit numbers, we can compute $1 \times 7^3 = 343$, then subtract the number of such base-8 integers from 1751_8 to 1777_8 (there are 21, namely $1751_8, \ldots, 1757_8, 1761_8, \ldots, 1767_8, 1771_8, \ldots, 1777_8$). This gives $343 - 21 = 322$ four-digit numbers.

    Thus, the desired answer is $1000 - 7 - 49 - 343 - 322 = \boxed{279}$. □

20. Note that R_1 is the circumradius of $\triangle ADC$ with side lengths b, $c/2$, and $c/2$, and R_2 is the circumradius of $\triangle BCD$ with side lengths a, $c/2$, and $c/2$.

    Using the area formula $[\triangle ABC] = \frac{abc}{4R}$, we obtain

$$[\triangle ACD] = [\triangle BCD] = \frac{b(c/2)^2}{4R_1} = \frac{a(c/2)^2}{4R_2}$$

    Simplifying the above equation yields $\frac{b}{R_1} = \frac{a}{R_2}$, or $bR_2 = aR_1$. From this, we obtain $\frac{R_1}{R_2} = \boxed{\frac{b}{a}}$. □

21. Let $a = \frac{x}{y} + \frac{y}{x}$. Then

$$a^2 = \frac{x^2}{y^2} + \frac{y^2}{x^2} + 2.$$

    Using the given equation we have

$$28 = a + (a^2 - 2) = a^2 + a - 2$$

    This quadratic equation has solutions $a = 5$ and $a = -6$. Since x and y are positive numbers, $a = 5$. Then

$$\frac{(x+y)^2}{xy} = \frac{x^2 + y^2 + 2xy}{xy} = \frac{x}{y} + \frac{y}{x} + 2 = a + 2 = \boxed{7}.$$

    □

22. It is obvious that $7d$ is larger than all $3a, 4b$ and $5c$, otherwise there is no positive integer solutions.

    Note that all numbers $3a, 4b$ and $5c$ must be equal. It can be proven by contradiction: assume that one of them is less than others, say i, then

$$3^{3a-i} + 3^{4b-i} + 3^{5c-i} = 3^{7d-i}.$$

However, left hand side is either 1 modulo 3 or 2 modulo 3, while the right hand side is divisible by 3. Thus we have a contradiction.

Now we have

$$3a = 4b = 5c.$$

Since $\text{lcm}(3,4,5) = 60$, $a = 20k, b = 15k, c = 12k$ for some $k \in \mathbb{N}$. Moreover $7d = 60k+1$. The minimum value of k satisfying the last equation is $k = 5$. Thus $a + b + c + d = 100 + 75 + 60 + 43 = \boxed{278}$. □

23. The first equation tells us that n is divisible by 10 but not 20, and the second equation tells us that $n + 1$ is divisible by 7 but not 21. Hence $n \equiv 10 \pmod{20}$ and $n \equiv 6, 13 \pmod{21}$. By Chinese remainder theorem, if we know the value of $n \pmod{20}$ and $n \pmod{21}$, we can uniquely find $n \pmod{420}$, which in this case is $n \equiv 90 \pmod{420}$ or $n \equiv 370 \pmod{420}$. The largest three-digit integer satisfying either of these congruences is $420 \times 2 + 90 = 930$, so the sum of the digits is $9 + 3 + 0 = \boxed{12}$. □

24. We observe that if $n > 0$ is good, and n is not a perfect square, then $-n$ is good. This follows as there exist positive $a, b \in S$ with $a \neq b$ and $n = ab$. Then $-n = (-a)b$, so $-n$ is good. Hence, in the desired sum, every good integer that is not a perfect square is canceled by its additive inverse, so we are only interested in perfect squares.

Every negative perfect square ($n = -k^2$, $k \in \{1, 2, \dots, 20\}$) is good since $n = (-k)(k)$. The sum of the negative perfect squares from 1^2 to 20^2 inclusive is $-(1^2 + 2^2 + \dots + 20^2) = -\frac{20 \times 21 \times 41}{6} = -2870$. However, some positive perfect squares are good: $2^2 = 1 \times 4$, $3^2 = 1 \times 9$, $4^2 = 2 \times 8$, $6^2 = 4 \times 9$, $8^2 = 4 \times 16$, $10^2 = 5 \times 20$, and $12^2 = 9 \times 16$. Hence the desired sum is $-2870 + (2^2 + 3^2 + 4^2 + 6^2 + 8^2 + 10^2 + 12^2) = \boxed{-2497}$. □

25. Consider the following diagram:

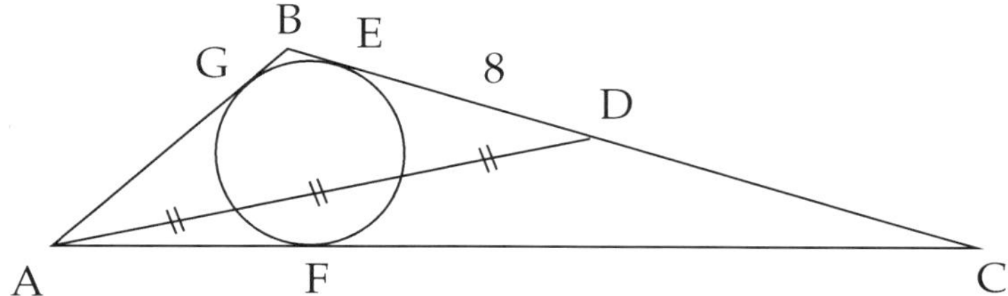

Let $AD = 3k$, so that each of the three equal segments has length k. By Power of a Point on D, we observe that $8^2 = k(2k) = 2k^2$, so $k = 4\sqrt{2}$ and $AD = 12\sqrt{2}$. Further, by Power of a Point on A, we see that $AG^2 = k(2k) = 8^2$, so $AF = AG = 8$.

Let $BE = x$; since $\overline{BG}$ and $\overline{BE}$ are both tangent, BG also equals x. Since D is the midpoint of $\overline{BC}$, $DC = x + 8$, so $EC = x + 16$ and $AC = 8 + (x + 16) = x + 24$. Applying Stewart's theorem on $\triangle ABC$ with cevian $\overline{AD}$:

$$(x+8)^2(x+8) + (x+24)^2(x+8) = (2x+16)(288 + (x+8)^2)$$

Divide both sides by $x + 8$:

$$(x+8)^2 + (x+24)^2 = 2(288 + (x+8)^2)$$

Letting $z = x + 8$, we have $z^2 + (z + 16)^2 = 576 + 2z^2 \iff 32z + 256 = 576 \iff z = 10$, so $x = z - 8 = 2$. Therefore the sides of the triangle are $AB = 10$, $BC = 20$, and $AC = 26$.

To find the radius of ω, we use Heron's formula along with the area formula $[\triangle ABC] = rs$ where s is the semiperimeter. The area of $\triangle ABC$ is (with $s = 28$):

$$[\triangle ABC] = \sqrt{28(18)(8)(2)} = 24\sqrt{14}$$

Then $24\sqrt{14} = 28r$, or $r = \frac{6\sqrt{14}}{7}$. The area of ω is $\pi r^2 = \boxed{\frac{72}{7}\pi}$.

As a side note, it can be checked that if $\triangle ABC$ has $AB = 10$, $BC = 20$, and $AC = 26$, then median $\overline{AD}$ is indeed trisected by ω. □

1. The difference is $(1+3+5+...+199)-(1+2+3+...+100) = (1-1)+(3-2)+(5-3)+...+(199-100) = 1+2+3+...+99 = \frac{99\times100}{2} = \boxed{4950}$. □

2. Note that exactly one of p, $p+10$, and $p+20$ is divisible by 3. We see that $p=3$ works as 13 and 23 are prime, but no other primes p work, as either $p+10$ or $p+20$ is a multiple of 3 larger than 3. Hence the answer is $\boxed{1}$. □

3. No matter what the first chip drawn is, there are 9 remaining chips, and 4 of them have the same parity (odd/even) as the first chip, in which the sum will be even. The probability is $\boxed{\frac{4}{9}}$. □

4. Extend AB and CD past B and C to meet at point P, as shown:

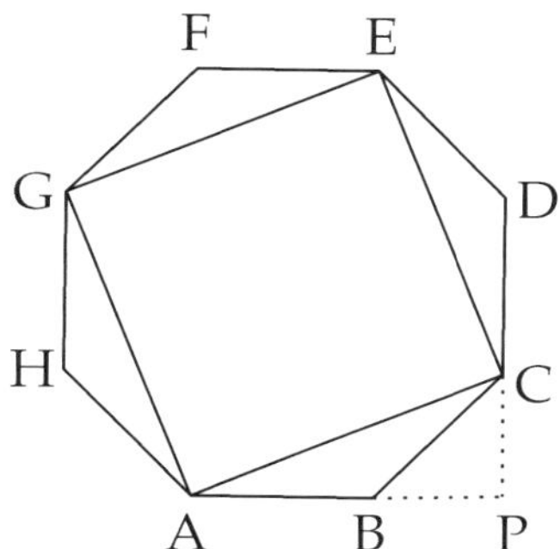

We have $AB = BC = 2$, and since $\angle P = 90^\circ$, it follows $BP = PC = \sqrt{2}$. Then $AP = 2+\sqrt{2}$ and $PC = \sqrt{2}$, so by the Pythagorean theorem, $AC^2 = AP^2 + PC^2 = (2+\sqrt{2})^2 + (\sqrt{2})^2 = (6+4\sqrt{2})+2 = \boxed{8+4\sqrt{2}}$.

Alternate solution: Note that $\angle B = 135^\circ$. By the law of cosines on $\triangle ABC$:

$$[ACEG] = AC^2 = 2^2 + 2^2 - 2\cdot 2^2 \cos 135^\circ = 8 - 8\left(-\frac{\sqrt{2}}{2}\right) = \boxed{8+4\sqrt{2}}.$$

□

5. Let D be the distance traveled in his morning commute, so that $2D$ is the distance traveled in his afternoon commute. Let h be the number of hours spent in his morning commute. We have the following equations:

$$D = 45h$$
$$2D = 60(1.5-h)$$

Then $2D = 90h = 60(1.5-h) = 90-60h$, so $150h = 90 \implies h = \frac{3}{5}$. Therefore he spent $\frac{3}{5}$ hours, or $\boxed{36}$ minutes, in his morning commute. □

6. As $\overline{MM} = 11\times M$, we see that $\overline{M2M2M}$ is divisible by 11. Using the divisibility rule for 11, we have that $3M-4 \equiv 0 \pmod{11}$, so the only possibility is $M=5$. Checking, we see that 52525 is divisible by 11 and 5, so it is divisible by 55, and $M = \boxed{5}$. □

7. He rolls the die exactly three times if and only if the sum of his first two rolls is 5 or less, and the sum of his first three rolls is 6 or greater. There are $6^3 = 216$ equally likely outcomes for David's first three rolls (even if he stops rolling prior). We do casework on David's 3rd roll.

   If David's third roll is 4, 5, or 6, then any outcome whose first two rolls sum to less than 6 is valid, namely $(1,1)$, $(1,2)$, $(1,3)$, $(1,4)$, $(2,1)$, $(2,2)$, $(2,3)$, $(3,1)$, $(3,2)$, $(4,1)$ ($10 \times 3 = 30$ outcomes).

   If David's third roll is 3, then any outcome whose first two rolls sum to at least 3 but less than 6 is valid, giving $9 \times 1 = 9$ outcomes.

   If David's third roll is 2, then any outcome whose first two rolls sum to 4 or 5 is valid, giving 7 outcomes.

   If David's third roll is 1, then any outcome whose first two rolls sum to 5 is valid, giving 4 outcomes.

   Altogether, the probability is $\frac{30+9+7+4}{216} = \frac{50}{216} = \boxed{\frac{25}{108}}$. □

8. Reflect the third dotted segment across the shorter side, as shown:

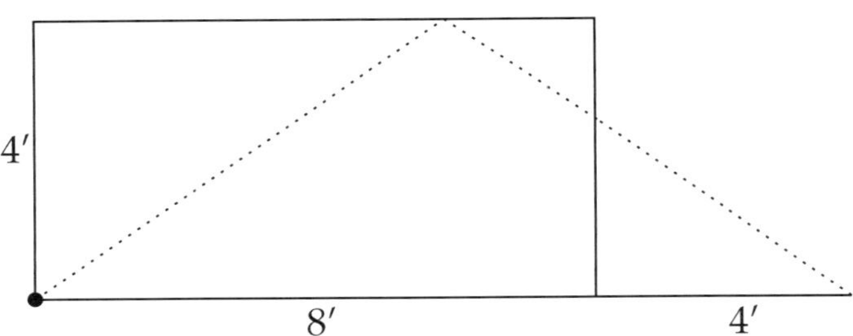

   The total length the ball travels is twice the length of one of the dotted segments. By the Pythagorean theorem, the length of one of the dotted segments is $\sqrt{4^2 + 6^2} = 2\sqrt{13}$ ft, so the total length the ball travels is $2 \times 2\sqrt{13} = 4\sqrt{13}$ ft. Since $\sqrt{13} \approx 3.6$, the total length the ball travels is approximately $\boxed{14.4}$ feet. □

9. Since the three numbers are in arithmetic progression, the middle number is the average of the first and last numbers, so $\log_2 2020 = \frac{\log_2 a + \log_2 b}{2}$, or $2\log_2 2020 = \log_2 a + \log_2 b = \log_2 ab$. Since $2\log_2 2020 = \log_2 2020^2$, we have $\log_2 2020^2 = \log_2 ab \implies ab = \boxed{2020^2}$. □

10. We can compute $10! = 3628800$, then long divide by 99 to get the remainder. However we give a faster and more interesting solution.

    We can find $10! \pmod 9$ and $\pmod{11}$, then the Chinese remainder theorem implies that we can find $10! \pmod{99}$. Clearly, $10! \equiv 0 \pmod 9$. Using either Wilson's theorem or the divisibility rule for 11, we establish that $10! \equiv 10 \pmod{11}$. Hence we are looking for a number which is $0 \pmod 9$ and $10 \pmod{11}$, which is $\boxed{54}$. □

11. We want the number of positive integer solutions to $a + b + c + d = 20$, where $a, b, c, d \geq 1$ and $b = d$.

If $b = d = 1$, then this is equivalent to finding the number of positive integer solutions to $a + c = 18$, which is 17. If $b = d = 2$, then this is equivalent to finding the number of positive integer solutions to $a + c = 16$, or 15 solutions. More generally, if $b = d = i$, then the number of solutions is $19 - 2i$. Altogether we get $17 + 15 + 13 + ... + 1 = 9^2 = \boxed{81}$. □

12. Let point P be on $\overline{AB}$ such that BCDP is a parallelogram:

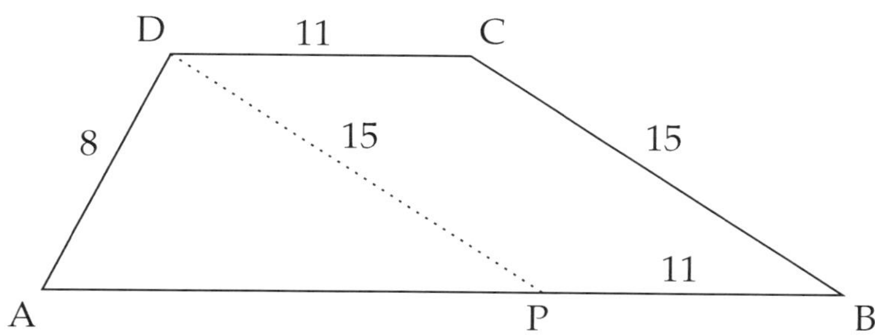

We have $m\angle BCD + m\angle CDP = 180°$; since it is given that $\angle C + \angle D + 270°$, it follows that $\angle PDA = 90°$, so $\triangle ADP$ is an 8-15-17 triangle. Then $AB = 17 + 11 = \boxed{28}$. □

13. We see that $\underbrace{2019^9 + ... + 2019^9}_{m \text{ times}} = (2019^{19})^2 = 2019^{38}$, so $m \times 2019^9 = 2019^{38}$, giving $m = \boxed{2019^{29}}$. □

14. Using the fact that $\gcd(a, b) = \gcd(b, a - b)$ for integers $a \geq b$, we have $\gcd(6n + 15, 10n + 21) = \gcd(6n + 15, 4n + 6) = \gcd(4n + 6, 2n + 9) = \gcd(2n + 9, 2n - 3)$. The numbers $2n + 9$ and $2n - 3$ differ by 12, so the gcd must be a divisor of 12. The number 12 has six divisors (1, 2, 3, 4, 6, 12), so there are 6 candidates for the gcd.

    We notice that $6n + 15$ and $10n + 21$ are both odd, so the gcd must be odd, and only 1 and 3 are candidates. $\gcd(6n+15, 10n+21) = 1$ if $n = 1$ for example, and $\gcd(6n+15, 10n+21) = 3$ if $n = 3$ for example. Therefore there are $\boxed{2}$ possible values for the gcd of $6n + 15$ and $10n + 21$. □

15. There are three possible pairs of dials which can be rotated on any move (#1 and #2, #2 and #3, or #3 and #4). Observe that the order of which the dials are rotated is irrelevant, and that rotating a pair of dials $4 + n$ times is equivalent to rotating that pair n times.

    Hence, we can assume that dials #1 and #2 are rotated a times, dials #2 and #3 are rotated b times, and dials #3 and #4 are rotated c times, where $0 \leq a, b, c \leq 3$. There are $4^3 = 64$ possibilities for the triple (a, b, c); moreover, each triple gives a unique configuration. The number of configurations is $\boxed{64}$. □

16. Consider the following diagram:

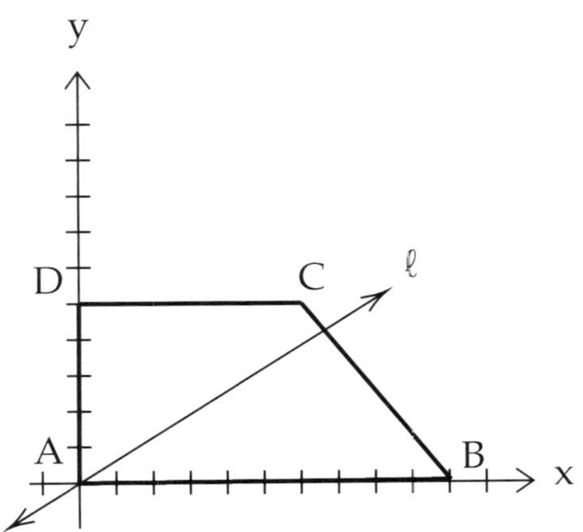

The area of trapezoid ABCD is $\frac{1}{2}(10+6)(5) = 40$, so the area of both regions must be 20. Note that $[\triangle ABC]$ is 25, so ℓ must intersect side $\overline{BC}$ (as opposed to $\overline{CD}$).

Since $AB = 10$, the height of the triangle measured from base AB should be 4. Therefore we wish to find the point on the segment $\overline{BC}$ whose y-coordinate is 4. This is the point $(\frac{34}{5}, 4)$. The slope of line ℓ is $\frac{4}{\frac{34}{5}} = \boxed{\frac{10}{17}}$. □

17. Rewrite the equation as $(2^x)^2 + 1 = 2^3 \times 2^x + 1$, or $(2^x)^2 - 8(2^x) + 1 = 0$. This is a quadratic in 2^x.

    Both solutions of the quadratic $y^2 - 8y + 1$ are positive, and the discriminant is positive, so there exist exactly two different real solutions x_1, x_2 for the original equation. To find their sum, we will first find the product of the solutions. By Vieta's formulas (or by the quadratic formula), the product of the two roots of $y^2 - 8y + 1$ is 1, so $2^{x_1} \times 2^{x_2} = 1 \implies 2^{x_1+x_2} = 1 \implies x_1 + x_2 = \boxed{0}$. □

18. Observe that $s(n) \equiv n \pmod 9$, so $s(n) = s(2n) \implies n \equiv 2n \pmod 9 \implies n \equiv 0 \pmod 9$, or in other words, n is a multiple of 9.

    Consider the addition $n + n$. If no carrying is performed when computing $n + n$, then $s(2n) = 2s(n)$, and $s(2n)$ cannot equal $s(n)$.

    If exactly one carry is performed, then $s(2n) = 2s(n) - 9$; in this case, $s(n) = s(2n)$ implies $s(n) = 2s(n) - 9$, or $s(n) = 9$. In this case, exactly one digit of n is 5 or greater. Listing out the possibilities by order of largest digit, we obtain 540, 531, 522, 630, 621, 720, 711, 810, 900 as well as permutations of these digits, which yields $4+6+3+4+6+4+3+4+1 = 35$ numbers.

    If exactly two carries are performed, then $s(2n) = 2s(n) - 18$, implying $s(n) = 18$. Here, exactly two of the three digits must be 5 or greater, so we have 774, 864, 873, 882, 954, 963, 972, 981, 990 as well as permutations of these digits, which yields $3+6+6+3+6+6+6+6+2 = 44$ numbers.

    If exactly three carries are performed, then $s(2n) = 2s(n) - 27$, or $s(n) = 27$. The only possibility is 999, which works as $s(1998) = s(999) = 27$.

    Hence the answer is $35 + 44 + 1 = \boxed{80}$. □

19. Partition the 101 coins into 3 groups A, B, C of sizes 50, 50 and 1 respectively, then compare the weights of A and B. If their weights are equal, each of A and B contains a counterfeit coin (while the other counterfeit coin is in C). Partition A into two subgroups A_1 and A_2 of sizes 25 and 25 and compare their weights. Their weights cannot be equal since A has exactly one counterfeit coin; moreover, the heavier subgroup contains no counterfeit coin. In total, the balance scale was used twice.

On the other hand, if the weights of A and B are not equal, then suppose WLOG that group A is heavier. We have two cases: either A contains one counterfeit coin while B contains two counterfeit coins, or A contains no counterfeit coins and B contains at least one counterfeit coin. In either case, partition A into two subgroups of sizes 25 and 25, and compare the weights of A_1 and A_2. The heavier group among A_1 and A_2 cannot contain any counterfeit coin. In this case, we have also identified a set of 25 real coins using the two-pan balance scale twice. Hence we can achieve this task with $\boxed{2}$ weighings. □

20. The trick is to construct D' on $\overleftrightarrow{BC}$ such that B is the midpoint of $D'D$:

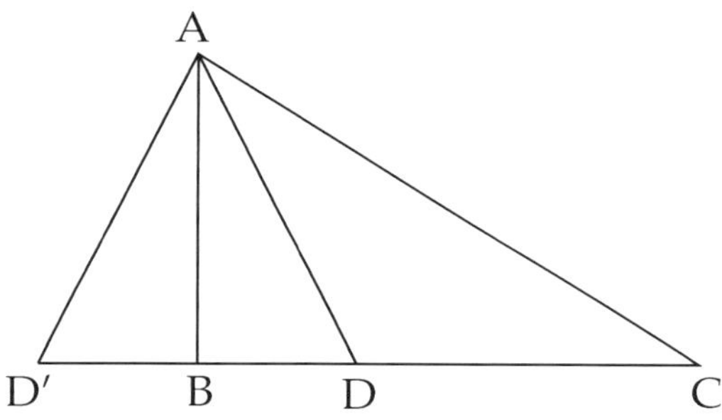

By construction, $2BD = D'D$. Also, triangle $AD'D$ is isosceles with $AD' = AD$, as altitude AB bisects $D'D$. The given condition implies $AD' \cdot DC = D'D \cdot AC$, or $\frac{AD'}{D'D} = \frac{AC}{DC}$, so AD is an angle bisector of $\angle CAD'$. Then $m\angle BAD = m\angle BAD' = \alpha$ for some α, and $m\angle DAC = 2\alpha$, so $\frac{m\angle DAC}{m\angle BAD} = \frac{2\alpha}{\alpha} = \boxed{2}$.

Alternate solution (Fake-solve): The desired ratio is presumably constant for all right triangles $\triangle ABC$ and points D that satisfy this property, so suppose $\angle BAC \approx 90^\circ$ and $\angle BCA \approx 0^\circ$. Then $DC \approx AC$, and $AD \approx 2BD$, so $\triangle ABD$ approaches a 30-60-90 triangle with $\angle BAD \approx 30^\circ$. As $\angle BAC \approx 90^\circ$, this implies $\angle DAC \approx 60^\circ$, giving the desired ratio of 2 as before. □

21. Note that there are two possible solutions (X, Y, Z) in positive integers: $(8, 3, 1)$ and $(7, 4, 3)$ (up to ordering). In either case, $\overline{XX}^2 + \overline{YY}^2 + \overline{ZZ}^2 = 11^2(X^2 + Y^2 + Z^2) = 11^2 \times 74 = 2 \times 11^2 \times 37$, and the number of divisors is $(1+1)(2+1)(1+1) = \boxed{12}$. □

22. Intuitively, M and N should be equidistant from C, as shown in the following diagram:

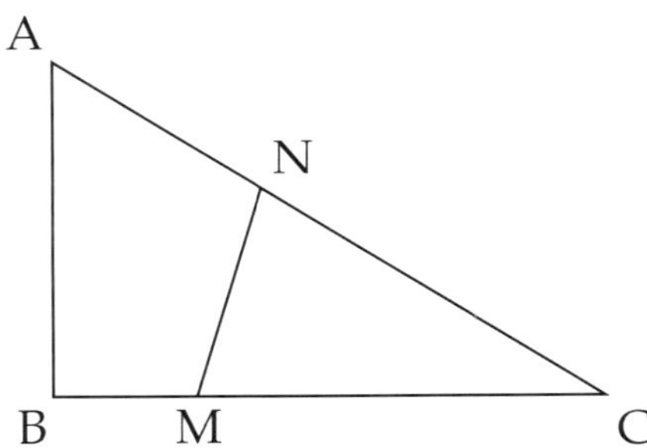

To prove this more rigorously, let $MC = x$ and $NC = y$. The area of $\triangle ABC$ is 6, so the area of $\triangle MNC$ must be 3. By the law of cosines, the length of $\overline{MN}$ satisfies

$$MN^2 = x^2 + y^2 - 2xy \cos \angle C = x^2 + y^2 - \frac{8}{5}xy$$

Therefore we wish to minimize MN^2 subject to the constraint that $[\triangle MNC]$ is 3, or equivalently:

$$\frac{1}{2}xy \sin \angle C = 3 \iff xy = 10$$

Then MN^2 can equivalently be written as $x^2 + y^2 - \frac{8}{5}(10) = x^2 + y^2 - 16$. For fixed xy, $x^2 + y^2$ is minimized when $x = y$; this can be shown with AM-GM or applying the fact that $(x - y)^2 \geq 0 \iff x^2 + y^2 \geq 2xy$.

Hence we have $x^2 + y^2 \geq 2xy = 20$, so $MN^2 \geq 20 - 16 = 4$ with equality when $x = y$. The minimum possible value of MN is $\boxed{2}$. □

23. It is easy to see that 1, as well as every prime number, appears in this sequence (ignore the "less than 1,000" condition for now). We now characterize the composite numbers that appear in this sequence. Observe the powers of 2 that appear in this sequence: 2, 4, 16. Note that 16 appears in this sequence but not 8, as $8 = 2 \times 4$, but 16 cannot be expressed as a product of previous terms. The number 32 does not appear in this sequence, since $32 = 16 \times 2$. Similarly, 64 and 128 do not appear in this sequence, as they can be expressed as products of previous terms already on the blackboard ($64 = 16 \times 4$, $128 = 16 \times 4 \times 2$). The next power of 2 to appear in this sequence is $2^8 = 256$. The next power of 2 to appear in this sequence is $2^{16} = 65536$. We extend this to all prime powers, in the following claim:

Claim: All numbers of the form p^{2^n}, where p is prime and $n \geq 0$, appear in this sequence.

Proof. We can proceed by induction on n, for some fixed p. Clearly, all primes appear in this sequence, so the statement is true for $n = 0$. Now, suppose for some $k \geq 0$, the numbers $p^{2^0}, \ldots, p^{2^k}$ appear in this sequence. Then it is not too hard to see that all of the numbers p^{2^i} where $0 \leq i \leq 2^{k+1} - 1$ can be expressed as a product of previous powers of p, and do not appear in the sequence (by considering the binary representations of 0, 1, $\ldots$, $2^{k+1} - 1$; for example, $5 = 101_2$, so $p^5 = p^4 \cdot p^1$). However, $p^{2^{k+1}}$ cannot be expressed as such, so it appears in the sequence. □

A corollary of the above claim is that, if this sequence is continued indefinitely, then every prime power can be expressed as a product of one or more distinct terms in this sequence.

Lastly, we show that if an integer contains two or more different primes in its prime factorization (such as 6), then it does not appear in this sequence. Consider a number of the form $p_1^{a_1}p_2^{a_2}\ldots p_j^{a_j}$ where $j \geq 2$, and $a_i \geq 1$. By the corollary above, the numbers $p_1^{a_1}$, $p_2^{a_2}, \ldots, p_j^{a_j}$ are all expressible as previous terms in the sequence. Combining these products, $p_1^{a_1}\ldots p_j^{a_j}$ is also expressible as a product of previous terms, and cannot be in the sequence.

Hence we have shown that all composite numbers in John's sequence are of the form p^{2^n} where p is prime and $n \geq 1$. The composite numbers in John's sequence less than 1000 are: 2^2, 2^4, 2^8, 3^2, 3^4, 5^2, 5^4, 7^2, 11^2, 13^2, 17^2, 19^2, 23^2, 29^2, and 31^2, or 15 terms.

Therefore, John's sequence contains the number 1, as well as 168 prime numbers and 15 composite numbers, or $1 + 168 + 15 = \boxed{184}$ numbers. □

24. Let $P(x) = (x^3 - 20x^2 - 20x + 1)^{10}$, let E be the sum of the even-indexed coefficients of $P(x)$ (i.e. the desired sum) and let $O = a_{29} + a_{27} + \ldots + a_1$ be the sum of the odd-indexed coefficients.

    Observe that -1 is a root of $x^3 - 20x^2 - 20x + 1$, so -1 is a root of $P(x)$, and $a_{30} - a_{29} + a_{28} - \ldots - a_1 + a_0 = 0 \implies E = O$. The sum of all coefficients is $P(1) = (-38)^{10} = 38^{10}$. Then $E = O = \frac{P(1)}{2} = \boxed{2^9 19^{10}}$. □

25. It may be tempting to claim that the expected number of hops is 4 as the frog hops an expected 1.5 units on each hop, but this is not quite true; if the frog is on "5", then it only hops 1 unit.

    For $n = 5, 4, 3, \ldots, 0$, let E_n denote the expected number of hops the frog needs to reach the number "6" given that the frog is already on the number n. Thus we have $E_5 = 1$ and $E_4 = \frac{1}{2}(1) + \frac{1}{2}(2) = \frac{3}{2}$. To find E_n where $n \leq 3$, we observe that with probability $\frac{1}{2}$, the frog hops forward 1 unit and makes expected $1 + E_{n+1}$, and with probability $\frac{1}{2}$, the frog hops forward 2 units and makes expected $1 + E_{n+2}$. Hence we have $E_n = \frac{1}{2}(1 + E_{n+1}) + \frac{1}{2}(1 + E_{n+2}) = 1 + \frac{1}{2}\left(E_{n+1} + E_{n+2}\right)$. We compute E_3, E_2, E_1, E_0 as such:

n	E_n
6	0
5	1
4	$\frac{3}{2}$
3	$1 + \frac{1}{2}\left(\frac{3}{2} + 1\right) = \frac{9}{4}$
2	$1 + \frac{1}{2}\left(\frac{9}{4} + \frac{3}{2}\right) = \frac{23}{8}$
1	$1 + \frac{1}{2}\left(\frac{23}{8} + \frac{9}{4}\right) = \frac{55}{16}$
0	$1 + \frac{1}{2}\left(\frac{55}{16} + \frac{23}{8}\right) = \frac{133}{32}$

The expected number of hops is $\frac{133}{32}$, which is closest to $\boxed{4.2}$. □

Solutions for Practice Exam 7.

1. We factor 10^{2018} from both the numerator and denominator, so that the expression simplifies to $\dfrac{10^{2018}(10^2+1)}{10^{2018}(10^2-1)} = \dfrac{10^2+1}{10^2-1} = \boxed{\dfrac{101}{99}}$. ☐

2. The units digits of the powers of 3 cycle: $3^1 = 3$, $3^2 = 9$, $3^3 \equiv 7$, $3^4 \equiv 1$, and so on. So any number of the form 3^{4k} has units digit 1. Since 2020 is a multiple of 4, 3^{2020} has units digit 1. Similarly, 7^{4k} also has units digit 1, so 7^{2020} has units digit 1. The answer is $1+1 = \boxed{2}$. ☐

3. Label the couple as A_1 and A_2, and label the child as C. There are 3 ways to choose which seats the couple and child occupy (seats 1-3, 2-4, or 3-5). There are 2 ways to choose the ordering of the couple (either A_1CA_2 or A_2CA_1). The remaining 2 people can be seated in $2! = 2$ ways. Altogether, the number of ways is $3 \times 2 \times 2 = \boxed{12}$. ☐

4. Since the similarity ratio is $5:4$, the perimeter of the smaller triangle is $\frac{30\cdot 4}{5} = 24$. Let the length of the shortest side of the smaller triangle be a. Then the other lengths are $a+2$ and $a+4$. Since $a+(a+2)+(a+4) = 24$, we have $a = \boxed{6}$. ☐

5. Let $x > 0$ be such that $x - \frac{1}{x} = 2$. Squaring both sides yields $x^2 + \frac{1}{x^2} - 2 = 4 \implies x^2 + \frac{1}{x^2} = 6$. Hence we have $f(2) = \boxed{6}$.

   It helps to check that there is exactly one positive value of x which satisfies $x - \frac{1}{x} = 2$. Multiplying by x, this implies $x^2 - 1 = 2x \iff x^2 - 2x - 1 = 0$. By the quadratic formula, $x = \frac{2 \pm \sqrt{8}}{2} = 1 \pm \sqrt{2}$; since x is positive, x must equal $1 + \sqrt{2}$. An example function which satisfies the given condition is the function $f(x) = x^2 + 2$. ☐

6. We observe that n, $n+1$, and $n+2$ cannot be divisible by 4, as such a number cannot be the product of two primes, so $n \equiv 1 \pmod{4}$. We can test numbers $1 \pmod{4}$ greater than 100 to see that $n = 121$ is minimum, so the answer is $1+2+1 = \boxed{4}$. ☐

7. There are $\binom{10}{2} = 45$ equally likely outcomes. We subtract the number of outcomes in which the product is odd. Since there are 5 odd numbers, the number of such outcomes is simply $\binom{5}{2} = 10$. The desired probability is $\dfrac{45-10}{45} = \boxed{\dfrac{7}{9}}$. ☐

8. Consider the cross-section containing the altitude with the lateral face. With some careful visualization and 30-60-90 triangles, we obtain that the base is an equilateral triangle with side $4\sqrt{3}$:

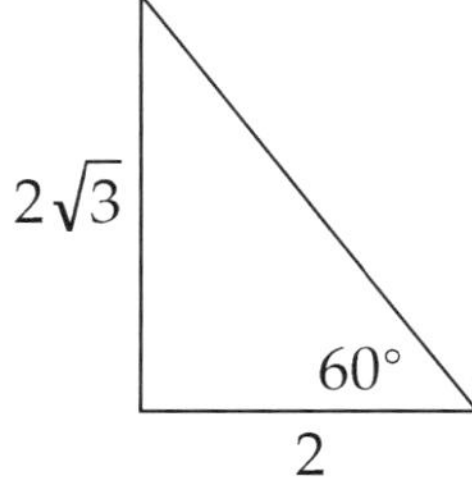

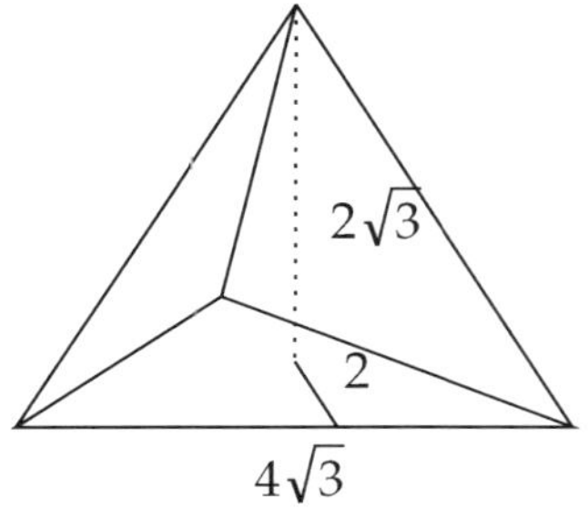

The base area is $\frac{(4\sqrt{3})^2\sqrt{3}}{4} = 12\sqrt{3}$, and the volume V is

$$V = \frac{1}{3}(12\sqrt{3})(2\sqrt{3}) = \boxed{24}.$$

□

9. We see that $1111 - 111 + 11 - 1 = 1010$ (13 keystrokes), so 2020 can be achieved using $\boxed{27}$ keystrokes: $1111 - 111 + 11 - 1 + 1111 - 111 + 11 - 1$. It can be checked that 27 is minimal (by observing that $1111 + 1111$ must appear in the sum, then finding the minimum number of keystrokes in order to obtain 202). □

10. Rewrite $n^2 + 8n - 85$ as $(n + 4)^2 - 101$. Then it is clear that 101 divides $n^2 + 8n - 85$ if and only if 101 divides $(n + 4)^2$, which occurs if and only if $n \equiv -4 \pmod{101}$. The candidate integers are $n = 97, 198, 299, 400, 501, 602, 703, 804, 905$, or $\boxed{9}$ values. □

11. Note that we can choose one directed segment (arrow) on the first "row" of the diagram, one directed segment on the second row, one on the third, and one on the fourth. This uniquely determines a path from A to B. The number of ways to choose directed segments in this way (and hence, the number of paths from A to B) is $2 \times 4 \times 6 \times 8 = \boxed{384}$. □

12. Let D be the foot of the altitude from A. Since $\triangle ABM$ is isosceles, $BD = DM$, so let $BD = DM = x$. Then $MC = 2x$. Applying the Pythagorean theorem on $\triangle ADM$ and $\triangle ADC$, we have

$$\begin{aligned} AD^2 + x^2 &= 36 \\ AD^2 + (3x)^2 &= 64 \end{aligned}$$

Subtracting the first equation from the second, we obtain $8x^2 = 28 \implies x = \frac{\sqrt{14}}{2}$. Then $BC = 4x = 2\sqrt{14}$ and the perimeter is $\boxed{14 + 2\sqrt{14}}$.

Alternate solution: Let $BM = MC = x$. Applying Stewart's theorem on $\triangle ABC$, we have

$$\begin{aligned} 8^2x + 6^2x &= 2x(6^2 + x^2) \\ 100x &= 2x(36 + x^2) \\ 50 &= 36 + x^2 \\ x &= \sqrt{14} \end{aligned}$$

Hence $BC = 2\sqrt{14}$ and we obtain the same answer as before. □

13. The desired region is a trapezoid, as shown:

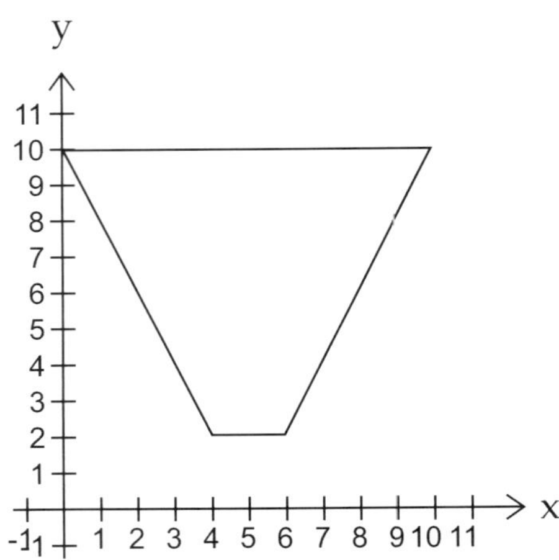

The vertices of the trapezoid are $(0,10)$, $(4,2)$, $(6,2)$, and $(10,10)$, so the parallel bases have lengths 10 and 2, and the height is 8. The area is $\frac{1}{2}(8)(10+2) = \boxed{48}$. □

14. Let x, y and z be such positive integers with $x+y+z = 314$. We want to find the maximum number of zeroes the product xyz could end with. Let $2^{x_1}5^{x_2}|x$, $2^{y_1}5^{y_2}|y$ and $2^{z_1}5^{z_2}|z$. One of x_2, y_2, z_2 must be 0, otherwise $5|314$, which is a contradiction. Say $x_2 = 0$. On the other hand, $y_2, z_2 \le 3$, otherwise $y, z \ge 625$, which is again a contradiction. So we can have $y = z = 125$ and $x = 64$. Then $xyz = 10^6$ and thus the answer is $\boxed{6}$. □

15. The prime factorization of $20!$ is $2^{18}5^4k$, where k is not divisible by 2 or 5. Note that we only care about the powers of 2 and 5 in the prime factorization, as we can choose a divisor randomly by choosing the powers of 2, 3, 5, ..., 19 randomly.

    Hence we can find the probability that a randomly chosen divisor of $2^{18}5^4$ is divisible by 100. This number has $(18+1)(4+1) = 95$ divisors, of which $17 \times 3 = 51$ are divisible by 100. The probability is $\boxed{\frac{51}{95}}$. □

16. Rotate $\triangle ABP$ 60° about B to form triangle CBP' as shown:

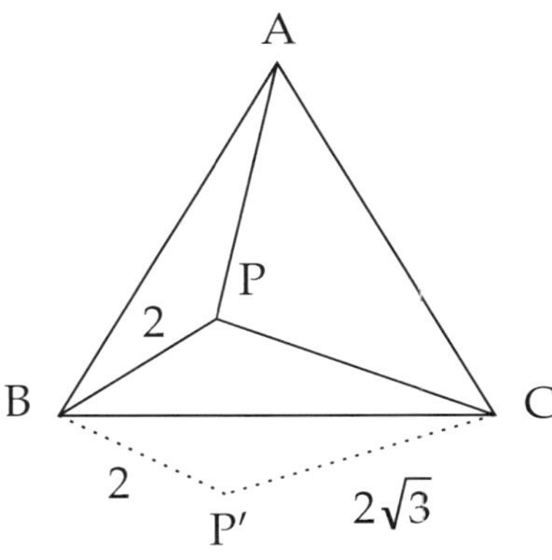

Note that $\angle PBP' = 60^\circ$. Since $BP = BP' = 2$, triangle PBP' is equilateral, so $PP' = 2$. Further, $\angle BP'C = 150^\circ$, so $\angle PP'C = 90^\circ$. By the Pythagorean theorem, $PC^2 = 2^2 + (2\sqrt{3})^2 = 16$, so $PC = \boxed{4}$. □

17. Rewrite $1 - \frac{1}{k^2}$ as $\frac{(k-1)(k+1)}{k^2}$. Then we have

$$\frac{1 \cdot 3}{2^2} \times \frac{2 \cdot 4}{3^2} \times \ldots \times \frac{(n-1)(n+1)}{n^2} = \frac{n+1}{16}$$

or, more concisely:

$$\frac{(n-1)!(n+1)!/2}{(n!)^2} = \frac{n+1}{16}$$

Simplifying the left-hand side, we obtain $\frac{n+1}{2n} = \frac{n+1}{16}$, or $2n = 16 \implies n = \boxed{8}$. □

18. Notice that $n = 1, 10, 100, 1000$ do not satisfy the given condition since $s(kn)$ can never equal 1 (unless k is a power of 10, which is forbidden). Hence we will assume n is not a power of 10.

We show that for all remaining positive integers n, there exists a k which is not a power of 10 such that $s(n) = s(kn)$. We prove this by considering two cases: i) The only prime factors of n are 2 and/or 5, and ii) n is divisible by some prime $p \neq 2, 5$.

For the first case, if the only prime factors of n are 2 and/or 5, then we can find a power of 10 which is divisible by n (specifically, if $n = 2^a 5^b$ where $a \neq b$, then take $10^{\max(a,b)}$). Then the number $N = \underbrace{11\ldots1}_{d(n)}\underbrace{00\ldots0}_{\max(a,b)}$ is a multiple of n whose sum of digits is $d(n)$, and the ratio $k = \frac{N}{n}$ is not a power of 10.

For the second case, let $n = 2^a 5^b M$ where M is not divisible by 2 or 5, and M is divisible by a prime $p \neq 2, 5$. Using the fact that $\gcd(M, 10) = 1$, there exists an integer q such that $10^q \equiv 1 \pmod{M}$. Consider the number $10^q + 10^{2q} + 10^{3q} + \ldots + 10^{qd(n)}$. This number has the decimal form $10\ldots010\ldots01\ldots1$, is divisible by M, and has sum of digits equal to $d(n)$. Pad this number using $\max(a, b)$ 0's to obtain a multiple of n with sum of digits equal to $d(n)$.

Hence all n except 1, 10, 100, 1000 work, so the answer is $1000 - 4 = \boxed{996}$. □

19. We use geometric probability. The sample space is the square $[0, 2] \times [0, 2]$ with area 4. Using the triangle inequality, the desired area is the region bounded by the inequalities $a + b > 1$, $a + 1 > b$, and $b + 1 > a$. Plotting these inequalities in the ab-plane, we can see that the desired area is $\frac{5}{8} \times 4 = \frac{5}{2}$, and the probability is $\frac{\frac{5}{2}}{4} = \boxed{\frac{5}{8}}$. □

20. We look at the change in surface area when one tetrahedron is sliced off. Initially, the visible surface area of the three (right isosceles) triangular faces of one of the tetrahedrons is $3 \times \frac{1}{4} = \frac{3}{4}$. After one tetrahedron is sliced off, we see that the visible surface area is an equilateral triangle of side length 1, whose area is $\frac{\sqrt{3}}{4}$. Then slicing off one tetrahedron changed the surface area by $\frac{\sqrt{3}}{4} - \frac{3}{4}$, or in other words, the surface area decreased by $\frac{3}{4} - \frac{\sqrt{3}}{4}$.

Since eight tetrahedra are sliced off, the surface area decreases by $6 - 2\sqrt{3}$. The original surface area is $6 \times 4^2 = 96$ square inches, so the surface area of the remaining solid is $96 - (6 - 2\sqrt{3}) = \boxed{90 + 2\sqrt{3}}$.

Alternate solution: We can compute the surface area directly. Each of the six octagonal faces can be thought of as a 4×4 square with four right isosceles triangles of side length $\frac{\sqrt{2}}{2}$ sliced off. Hence the area of each octagonal face is $4^2 - 4(\frac{1}{4}) = 15$, so the area of the octagonal faces is 90. The area of each triangular face is $\frac{\sqrt{3}}{4}$, and there are eight of them, so the area of the triangular faces is $2\sqrt{3}$. The surface area is $\boxed{90 + 2\sqrt{3}}$ as before. □

21. Note that we can assume $x, y \geq 0$; if x (or y) is negative, then replacing x with $-x$ yields a larger value for $x + y$.

We give a geometric interpretation for this problem. Consider two right triangles with side lengths x, 1, $\sqrt{x^2+1}$, and y, 2, $\sqrt{y^2+4}$ configured as shown:

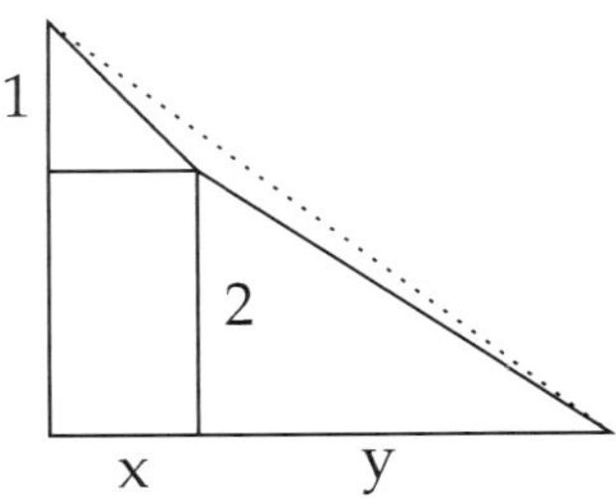

The sum of the lengths of the two hypotenuses is given to be 7, which implies that the hypotenuse of the large triangle (dotted) has length at most 7. This implies that

$$(x+y)^2 + (2+1)^2 \leq 7^2 \implies (x+y)^2 \leq 40 \implies x + y \leq 2\sqrt{10}$$

Equality can occur when the two hypotenuses coincide with the dotted segment. The maximum possible value of $x + y$ is therefore $\boxed{2\sqrt{10}}$. □

22. Since Qi has at least one of each type of coin, and since the number of dimes does not affect the average value of the coins (as dimes are worth 10 cents), we can assume that Qi has 1 dime.

Let p, n, q denote the number of pennies, nickels, and quarters, respectively. Then $\frac{p+5n+25q}{p+n+q} = 10 \iff p + 5n + 25q = 10p + 10n + 10q$. Looking at this equation modulo 5, we see that $p \equiv 0 \pmod{5}$. Suppose $p = 5$, so that $5 + 5n + 25q = 50 + 10n + 10q \iff 15q = 5n + 45 \iff 3q = n + 9$. The solution that minimizes $p + n + q$ is $n = 3$ and $q = 4$, giving $(p, n, d, q) = (5, 3, 1, 4)$, or 13 coins. If $p = 10$, then because she has at least one nickel, one dime, and one quarter, then she must have at least 13 coins no matter what. Hence the minimum possible number of coins is $\boxed{13}$. □

23. The area of the hexagon whose coordinates are $(x, 2^x)$, $(x+1, 2^{x+1})$, $(x+2, 2^{x+2})$, ..., $(x+5, 2^{x+5})$ is 144. Applying the shoelace formula, we have

$$\frac{1}{2}\left|4 \times 2^x - 2(2^{x+1} + 2^{x+2} + 2^{x+3} + 2^{x+4}) + 4 \times 2^{x+5}\right| = 144$$
$$\left|2 \times 2^x - (2^{x+1} + 2^{x+2} + 2^{x+3} + 2^{x+4}) + 2 \times 2^{x+5}\right| = 144$$

Factor out 2^x from the LHS:

$$\begin{aligned} 2^x|2-(2+4+8+16)+64| &= 144 \\ 36 \times 2^x &= 144 \\ 2^x &= 4 \end{aligned}$$

Hence $x = \boxed{2}$. □

24. By the quadratic formula, if $z^2+az+b=0$, then $z = \frac{-a\pm\sqrt{a^2-4b}}{2}$, so we wish to determine the set of all complex numbers z which can be written in this form, for some $a, b \in [0,1]$.

Note that if $a^2-4b \geq 0$, then z is real, and the range of possible *real* numbers z is the interval $[-1,0]$ (a^2-4b is at most 1, so $-a-\sqrt{a^2-4b} \geq -2$). Otherwise, $a^2-4b < 0$, in which $\frac{-a\pm\sqrt{a^2-4b}}{2} = -\frac{a}{2} \pm \frac{\sqrt{4b-a^2}}{2}i$. Geometrically, this represents the point $(-\frac{a}{2}, \pm\frac{\sqrt{4b-a^2}}{2})$. Consider some $a \in [0,1]$. The real part ("x-coordinate") is $-\frac{a}{2}$ and the imaginary part ("y-coordinate") is $\frac{\sqrt{4b-a^2}}{2}$, whose absolute value is at least 0 and at most $\frac{\sqrt{4-a^2}}{2}$ (when $b=1$). Hence for a fixed $a \in [0,1]$, the set of complex numbers in R which are the root of some polynomial of the form x^2+ax+b (for some b) is $-\frac{a}{2}+mi$ where $|m| \leq \frac{4-a^2}{2}$.

Consider points of the form $(-\frac{a}{2}, \frac{\sqrt{4-a^2}}{2})$, where $a \in [0,1]$. Notice that these points lie on the unit circle; this is checked as $\left(-\frac{a}{2}\right)^2 + \left(\frac{\sqrt{4-a^2}}{2}\right)^2 = \frac{a^2}{4}+\frac{4-a^2}{4} = 1$. This gives us enough information to plot the region R. Note that R also contains the interval $[-1,0]$, but this has area zero.

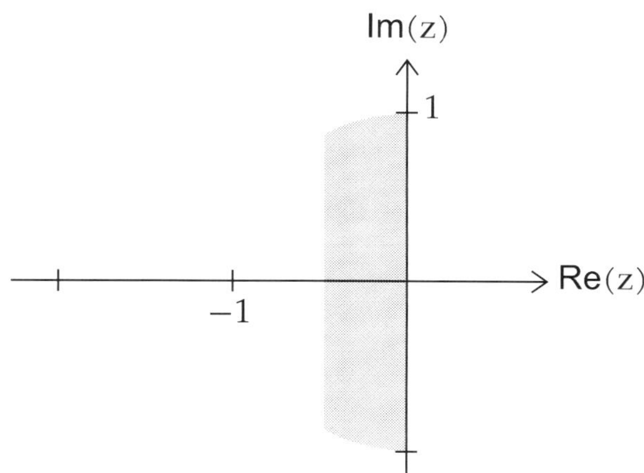

The area of R can be calculated with basic geometry. For instance, we can divide R into an isosceles triangle and two $30°$ sectors, as shown:

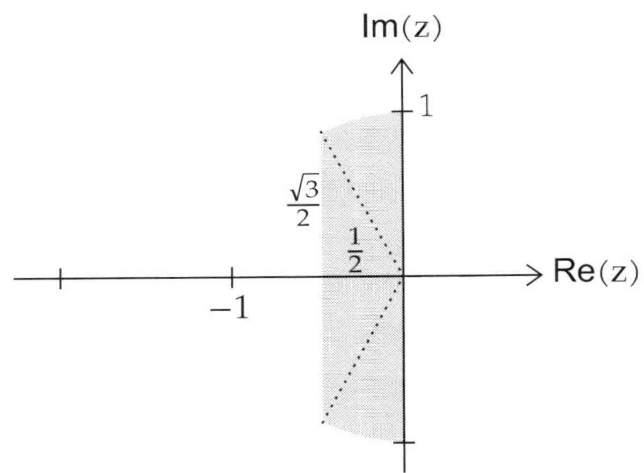

We see that the area of R is $\boxed{\frac{\sqrt{3}}{4}+\frac{\pi}{6}}$. □

25. Note that we only care about how many red marbles are in the bag, and not the actual colors themselves.

We will denote by $p_{t,n}$ the probability that, immediately before the tth marble selection, the bag contains exactly n red marbles. We have $(p_{1,1}, p_{2,1}, p_{3,1}, p_{4,1}, p_{5,1}) = (1, 0, 0, 0, 0)$ since the bag initially contains one red marble.

We compute $p_{2,n}$ and $p_{3,n}$, etc. recursively. Specifically, to find $p_{t,k}$, we see that at time t, the bag contains k red marbles if and only if the bag contained k red marbles at time $t-1$ and a red marble was selected (with probability $\frac{k}{5}$), or if the bag contained $k-1$ red marbles at time $t-1$ and a non-red marble was selected (with probability $1-\frac{k-1}{5}$). Thus we have the recurrence $p_{t,k} = p_{t-1,k}\frac{k}{5} + p_{t-1,k-1}(1-\frac{k-1}{5})$. Using this, we can compute $p_{2,n}$, $p_{3,n}$, etc. (which represent probability distributions):

$$(p_{2,1}, p_{2,2}, p_{2,3}, p_{2,4}, p_{2,5}) = \left(\frac{1}{5}, \frac{4}{5}, 0, 0, 0\right)$$
$$(p_{3,1}, p_{3,2}, p_{3,3}, p_{3,4}, p_{3,5}) = \left(\frac{1}{25}, \frac{12}{25}, \frac{12}{25}, 0, 0\right)$$
$$(p_{4,1}, p_{4,2}, p_{4,3}, p_{4,4}, p_{4,5}) = \left(\frac{1}{125}, \frac{28}{125}, \frac{72}{125}, \frac{24}{125}, 0\right)$$
$$(p_{5,1}, p_{5,2}, p_{5,3}, p_{5,4}, p_{5,5}) = \left(\frac{1}{625}, \frac{60}{625}, \frac{300}{625}, \frac{240}{625}, \frac{24}{625}\right)$$

Then immediately before the fifth marble is drawn, there is a $\frac{1}{625}$ chance that the bag contains 1 red marble, a $\frac{60}{625}$ chance that the bag contains 2 red marbles, and so on. The

probability that the fifth marble drawn is red is

$$\begin{aligned} p(\text{5th marble red}) &= \frac{1}{625} \cdot \frac{1}{5} + \frac{60}{625} \cdot \frac{2}{5} + \frac{300}{625} \cdot \frac{3}{5} + \frac{240}{625} \cdot \frac{4}{5} + \frac{24}{625} \cdot \frac{5}{5} \\ &= \frac{1 \cdot 1 + 60 \cdot 2 + 300 \cdot 3 + 240 \cdot 4 + 24 \cdot 5}{3125} \\ &= \frac{2101}{3125} \end{aligned}$$

Therefore $m + n = 2101 + 3125 = 5\boxed{226}$. □

Solutions for Practice Exam 8.

1. Initially, the water in the tomatoes weighs 90 lb, and the remaining mass weighs 10 lb. Afterwards, the remaining mass still weighs 10 lb, so in order for the tomatoes to be 80% water by weight, the water must weigh 40 lb. The tomatoes weigh $40 + 10 = \boxed{50}$ lb. □

2. Multiply all sides of the inequality by 2018 to obtain $\frac{2018}{4} \le n \le \frac{2018}{3}$, or equivalently $504.5 \le n \le 672.\overline{6}$. The integer solutions are $n = 505, 506, 507, \ldots, 672$, or $672 - 505 + 1 = \boxed{168}$ solutions. □

3. We can count how many monetary amounts Rachel can make which are multiples of \$5. The idea is that if Rachel can make $\$5k$, then she can also make $\$(5k + 1)$ and $\$(5k + 2)$ by using one or two \$1 bills.

   Using any combination of up to two each of the \$5, \$10, \$20 bills, Rachel can form all multiples of 5 between 0 and 70 dollars, inclusive (15 multiples of 5). Each multiple of 5 (e.g., $\$5k$) gives three different monetary amounts. Hence there are $15 \times 3 = 45$ different monetary amounts. However this counts \$0 as a possible amount, which is not allowed since one or more bills is needed, so the answer is $45 - 1 = \boxed{44}$. □

4. The radius of the steel ball is 6 inches, so its volume is $\frac{4}{3}\pi(6^3) = 288\pi$ cubic inches. If the volume of the water in the tank is V cubic inches, then the volume of the water and the steel ball is $V + 288\pi$ cubic inches. The height (water level) is obtained by dividing the volume by the base area, and the difference in water level can be obtained by dividing the volume of the ball by the base area.

   Since the base area is 8 square feet, or 1152 square inches, the change in the water level will be $\frac{288\pi \text{ in}^3}{1152 \text{ in}^2} = \frac{\pi}{4}$ inches. As $\pi \approx 3.14$, the change in water level is approximately $\boxed{0.8}$ inches. □

5. Note that if a is a positive integer, then $|\sqrt{a} - n| < \frac{1}{2}$ if and only if $-\frac{1}{2} < \sqrt{a} - n < \frac{1}{2}$, or $n - \frac{1}{2} < \sqrt{a} < n + \frac{1}{2}$. Squaring all sides of this inequality, we obtain $n^2 - n + \frac{1}{4} < a < n^2 + n + \frac{1}{4}$. Then the desired set is $\{n^2 - n + 1, \ldots, n^2 + n\}$, and the number of elements is $(n^2 + n) - (n^2 - n + 1) + 1 = \boxed{2n}$.

   Alternate solution: Fix a value of n, say $n = 2$. The set $\{a \in \mathbb{N} : |\sqrt{a} - 2| < \frac{1}{2}\}$ is the set $\{3, 4, 5, 6\}$ containing 4 elements. Plugging in $n = 2$ into the answer choices, only $\boxed{2n}$ works. □

6. We factor out $20!$, so that $N = 20!(1 + 21 + 21 \times 22 + 21 \times 22 \times 23)$. We can write the latter term as $22 + 21 \times 22 + 21 \times 22 \times 23 = 22(1 + 21 + 21 \times 23) = 22 \times 505 = 2 \times 5 \times 11 \times 101$. Hence $N = 20! \times 2 \times 5 \times 11 \times 101$, so $p = 101$ and the sum of its digits is $\boxed{2}$. □

7. The triangle is degenerate (area 0) if and only if the points $(a, |a|)$, $(b, |b|)$, $(c, |c|)$ lie on the same side of the graph of $y = |x|$, which occurs if and only if $a, b, c \ge 0$ or $a, b, c \le 0$. These two events each occur with probability $\frac{1}{8}$ and are disjoint (unless $a = b = c = 0$, which occurs with probability 0), so the desired probability is $1 - \frac{1}{8} - \frac{1}{8} = \boxed{\frac{3}{4}}$. □

8. Consider the diagram below, where P is the intersection of CD and $B'C'$ (note: a slightly different configuration arises if the square is rotated in the opposite direction, but the desired region is the same).

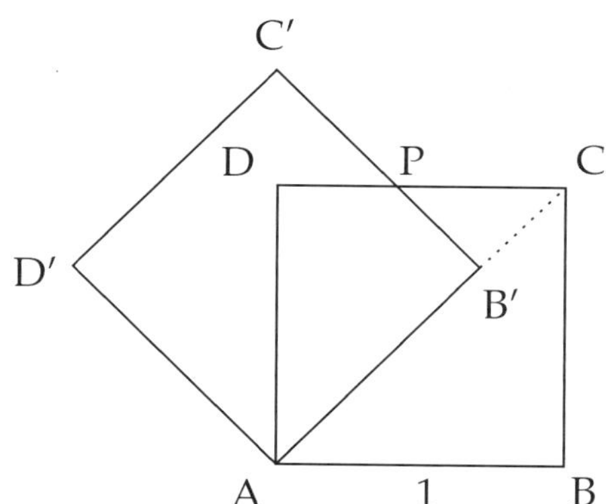

By some simple angle chasing, we have $\angle B'PD = 135^\circ \implies \angle B'PC = 45^\circ$, and $\angle B'CP = 45^\circ$, so $\triangle B'CP$ is an isosceles right triangle. To find the area of the overlap, we find the area of isosceles triangle ACD, then subtract the area of isosceles triangle $B'CP$. The area of $\triangle ACD$ is $\frac{1}{2}$. Further, we have $B'C = B'P = \sqrt{2} - 1$, so the area of $\triangle B'CP$ is $\frac{1}{2}\left(\sqrt{2}-1\right)^2 = \frac{3}{2} - \sqrt{2}$.

Therefore the area of $AB'PD$ is $\frac{1}{2} - \left(\frac{3}{2} - \sqrt{2}\right) = \boxed{\sqrt{2}-1}$. □

9. There are three cases: i) $x^2 - 3x + 1 = 1$. So $x = 3$ or $x = 0$. ii) $x^2 - 3x + 1 = -1$ but $x + 1$ is even. So $x = 1$. iii) $x + 1 = 0$ but $x^2 - 3x + 1 \neq 0$. So $x = -1$. Then there are $\boxed{4}$ solutions. □

10. Since gcd$(4, 25) = 1$, gcd$(6, 25) = 1$ and $\varphi(25) = 20$, we have $4^{20} \equiv 1 \pmod{25}$ and $6^{20} \equiv 1 \pmod{25}$ by Euler's theorem. Then,

$$4^{2018} + 6^{2018} \equiv 4^{18} + 6^{18} \pmod{25}.$$

By simple calculations we have $4^{18} \equiv 11 \pmod{25}$ and $6^{18} \equiv 16 \pmod{25}$. Thus $4^{2018} + 6^{2018} \equiv \boxed{2} \pmod{25}$. □

11. There are $9 \times 8 = 72$ equally likely outcomes representing the two chips Sean selects, if we consider order important. Let P denote their product. Using the definition of expectation:

$$\begin{aligned}
\mathbb{E}(P) &= \frac{1}{72}(1 \cdot 2 + 1 \cdot 3 + \ldots + 1 \cdot 9 + 2 \cdot 1 + 2 \cdot 3 + \ldots + 9 \cdot 8) \\
&= \frac{1}{72}\left((1 + 2 + 3 + \ldots + 9)^2 - (1^2 + 2^2 + \ldots + 9^2)\right) \\
&= \frac{1}{72}\left(45^2 - \frac{9 \times 10 \times 19}{6}\right) \\
&= \frac{145}{6} = \boxed{24\frac{1}{6}}
\end{aligned}$$

□

12. Let $m\angle BCA = \alpha$, so that $m\angle ABC = 2\alpha$. Let D be on $\overline{AC}$ such that BD bisects $\angle B$:

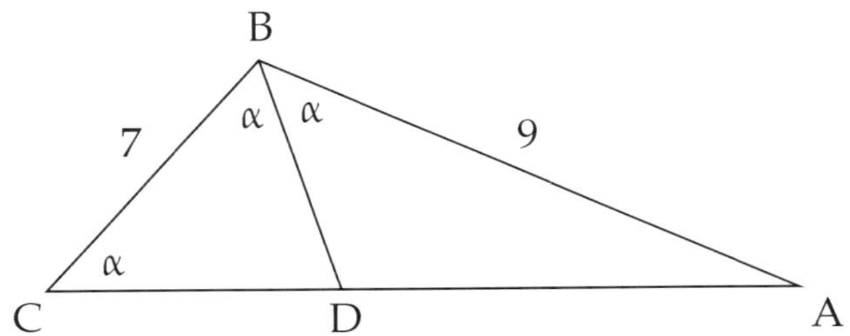

By the angle bisector theorem, $CD : DA = 7 : 9$, so let $CD = 7k$ and $DA = 9k$. Triangle BCD is isosceles, so $BD = 7k$.

Applying Stewart's theorem on $\triangle ABC$:

$$7^2(9k) + 9^2(7k) = 16k((7k)^2 + (7k)(9k))$$

With a bit of algebra, the above equation simplifies to $63 = 112k^2 \implies k^2 = \frac{63}{112} = \frac{9}{16}$, so $k = \frac{3}{4}$. The length of $\overline{AC}$ is $7k + 9k = 16k = \boxed{12}$. □

13. Note that $x \geq \frac{5}{4}$, otherwise $\sqrt{4x-5}$ is imaginary. Let $a = 4x - 5$, so that $6 - 4x = 1 - a$. Thus we wish to solve the equation $\sqrt{a} + \sqrt[3]{1-a} = 1$ where $a \geq 0$.

    Rewrite this equation as $\sqrt[3]{1-a} = 1 - \sqrt{a}$. Cube both sides to get $1 - a = 1 - 3\sqrt{a} + 3a - a\sqrt{a}$, or $a\sqrt{a} - 4a + 3\sqrt{a} = 0$. Note that $a = 0$ (i.e. $x = \frac{5}{4}$) is a solution; otherwise $a > 0$, and dividing by $\sqrt{a}$ gives $a - 4\sqrt{a} + 3 = 0$. This is a quadratic in terms of $\sqrt{a}$, and it factors to $(\sqrt{a} - 1)(\sqrt{a} - 3) = 0$, giving $\sqrt{a} = 1$ or $\sqrt{a} = 3$.

    Then $a = 0$, $a = 1$, and $a = 9$ are the solutions to the equation $\sqrt{a} + \sqrt[3]{1-a} = 1$, giving solutions $x = \frac{5}{4}, \frac{3}{2}$, and $\frac{7}{2}$. All three of these solutions work, so their sum is $\frac{5}{4} + \frac{3}{2} + \frac{7}{2} = \boxed{\frac{25}{4}}$. □

14. The last numbers counted by Bob and Charlie are 180 and 300, respectively.

    A positive integer n is counted by at least one person if and only if n is a multiple of 2, 3, or 5 less than or equal to 120, or n is a multiple of 3 or 5 less than or equal to 180, or n is a multiple of 5 less than 300. The number of integers n divisible by 2, 3, or 5 can be quickly counted and is equal to $120 - \varphi(120) = 120 - 32 = 88$, where $\varphi(\cdot)$ denotes Euler's totient function.

    The number of integers n greater than 120 and less than or equal to 180 which are divisible by 3 or 5 may be counted as follows: there are 60 numbers in the set $\{121, \ldots, 180\}$. For every 15 consecutive numbers, exactly $15 - \varphi(15) = 7$ numbers are divisible by 3 or 5. Then $7 \times 4 = 28$ numbers in the set $\{121, \ldots, 180\}$ are divisible by 3 or 5.

    Finally, we count the number of multiples of 5 greater than 180 and less than or equal to 300. This is the set $\{185, 190, \ldots, 300\}$, or 24 numbers.

    Thus the desired answer is $88 + 28 + 24 = \boxed{140}$. □

15. We use complementary counting, by finding the number of positive integers less than 10^4 satisfying $s(n) \leq 10$, then subtracting from $10^4 - 1$.

    We can specify each positive integer less than 10^4 uniquely using four digits $d_1d_2d_3d_4$ where leading zeros are allowed (e.g. 123 corresponds to 0123), and the d_i's are not all zero. Hence we seek the number of non-negative integer solutions to $d_1+d_2+d_3+d_4 \leq 10$ where $0 \leq d_i \leq 9$ and the d_i's are not all zero.

    Add a dummy variable $s := 10 - d_1 - d_2 - d_3 - d_4$. By stars and bars, the number of non-negative integer solutions to $d_1 + d_2 + d_3 + d_4 + s = 10$ with no other constraint is $\binom{14}{4} = 1001$. However this counts the illegal solution $(d_1, d_2, d_3, d_4, s) = (0, 0, 0, 0, 0, 10)$, as well as the four solutions where some d_i is equal to 10. Hence the number of valid solutions to $d_1 + ... + d_5 \leq 10$ is $1001 - 1 - 4 = 996$.

    By complementary counting, the number of such n is $10^4 - 1 - 996 = \boxed{9003}$. □

16. Note that the area of $\triangle ABC$, denoted $[\triangle ABC]$ is 12; this can be seen by drawing an altitude from A to BC.

    Let h_a, h_b, and h_c denote the lengths of the altitudes from P to sides BC, AC, and AB respectively. The area of $\triangle ABC$ is $[\triangle ABC] = [\triangle APB] + [\triangle APC] + [\triangle BPC] = 12$. We wish to maximize the product of these areas.

    By the AM-GM inequality, we have

    $$4 = \frac{[\triangle APB] + [\triangle APC] + [\triangle BPC]}{3} \geq \sqrt[3]{[\triangle APB][\triangle APC][\triangle BPC]}$$

    with equality if and only if the areas of triangles APB, APC, and BPC are equal. Hence, in order to maximize the product of the areas, the areas of each of these triangles must be $\frac{12}{3} = 4$. This implies the altitude from P to BC is 1, and that the foot of this altitude bisects BC. By the Pythagorean theorem, $PC = \sqrt{1^2 + 4^2} = \boxed{\sqrt{17}}$. □

17. Suppose y is fixed, so that the expression $x^2 + (y-4)x + (y^2 - 5y)$ is treated as a quadratic in x. For fixed y, the value of x which minimizes the expression is $x = -\frac{y-4}{2}$; substituting x with $-\frac{y-4}{2}$ yields $\frac{1}{4}(y-4)^2 - \frac{1}{2}(y-4)^2 + (y^2-5y) = (y^2-5y) - \frac{1}{4}(y-4)^2 = \frac{1}{4}(3y^2 - 12y - 16)$. Setting $y = 2$ (and $x = 1$) minimizes the entire expression, in which we obtain a minimum of $\boxed{-7}$. □

18. Since $p, q > 0$, we have that $p^2 + 3pq + q^2 > p^2 + 2pq + q^2 = (p+q)^2$, so $p^2 + 3pq + q^2$ is at least $(p+q+1)^2$.

    Note that $(p+q+1)^2 = p^2 + 2pq + q^2 + 2p + 2q + 1$. If $p^2 + 3pq + q^2 = (p+q+1)^2$, then $3pq = 2pq + 2p + 2q + 1$, or $pq - 2p - 2q = 1$. Adding 4 to both sides, $(p-2)(q-2) = 5$, giving $(p, q) = (7, 3), (3, 7)$.

    More generally, if $p^2 + 3pq + q^2 = (p+q+k)^2$ for some integer $k \geq 2$, then expanding the RHS and simplifying gives $3pq = 2pq + 2pk + 2qk + k^2$, or $pq - 2pk - 2qk = k^2$. Adding $4k^2$ to both sides, this simplifies to $(p-2k)(q-2k) = 5k^2$. We observe that if $k \geq 2$, there is no solution (p, q) in prime numbers; to prove this, let r be a prime divisor of k. Taking the equation modulo r, we have $(p-2k)(q-2k) \equiv 0 \pmod{r} \iff pq \equiv 0 \pmod{r}$. As r is prime, then by Euclid's lemma, r divides p or q. Further, observe that $p - 2k$ and

$q - 2k$ must be positive; otherwise they are both negative and the absolute value of their product is at most $4k^2 < 5k^2$. Hence either p or q is a multiple of r and greater than $2k$, implying that either p or q is composite. Hence there are no solutions for $k \geq 2$, so the only solutions are $(p, q) = (3, 7)$, $(7, 3)$, or $\boxed{2}$ ordered pairs. □

19. Let E denote the desired event, that Ben rolls at most three times. We consider the complementary event $\overline{E}$; that is, we can find 1 minus the probability that Ben rolls the die four or more times. Let X_n denote Ben's n^{th} roll, and let μ_n denote the average of Ben's first n rolls.

    The complementary event occurs if and only if $\mu_1, \mu_2, \mu_3 < 3.5$, or equivalently, $X_1 \leq 3$, $X_1 + X_2 \leq 6$, and $X_1 + X_2 + X_3 \leq 10$. To find $p(\overline{E})$, we can do cases on $X_1 + X_2$. There are $6^3 = 216$ equally likely outcomes from 3 dice rolls.

    If $X_1 + X_2 \leq 4$, then it is implied that $X_1 \leq 3$ and $X_1 + X_2 + X_3 \leq 10$. Hence $(X_1, X_2) = (1,1)$, $(1,2)$, $(2,1)$, $(2,2)$, $(1,3)$, or $(3,1)$, and X_3 can be anything, giving $6 \times 6 = 36$ outcomes.

    If $X_1 + X_2 = 5$, then $(X_1, X_2) = (1,4)$, $(2,3)$, $(3,2)$ (not $(4,1)$), and $X_3 \leq 5$, giving $3 \times 5 = 15$ outcomes.

    If $X_1 + X_2 = 6$, then $(X_1, X_2) = (1,4)$, $(2,3)$, $(3,2)$ and $X_3 \leq 4$, giving $3 \times 4 = 12$ outcomes.

    Hence, $p(\overline{E}) = \frac{36+15+12}{216} = \frac{7}{24}$, and $p(E) = 1 - p(\overline{E}) = 1 - \frac{7}{24} = \boxed{\frac{17}{24}}$. □

20. Let O_1 and O_2 be the centers of ω_1 and ω_2 respectively. Since $O_1O_2 = O_1A = O_2A$, triangle AO_1O_2 is equilateral.

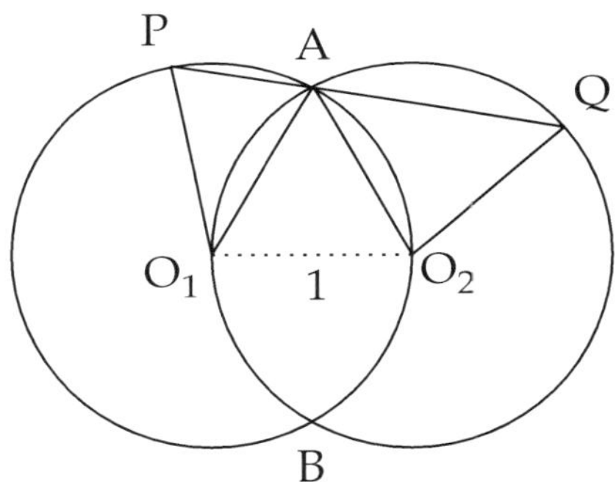

Let $\angle O_1PA = \angle O_1AP = \alpha$. With a little angle chasing, we see that $\angle O_2AQ = \angle O_2QA = 120° - \alpha$, $\angle AO_2Q = 2\alpha - 60°$, and $\angle PO_1A = 180° - 2\alpha$. Let $PA = x$ and $AQ = 2x$. Applying the law of cosines on $\triangle PO_1A$ and AO_2Q:

$$x^2 = 2 - 2\cos(180° - 2\alpha)$$
$$4x^2 = 2 - 2\cos(2\alpha - 60°)$$

Using basic trigonometric identities, we can write $\cos(180° - 2\alpha)$ as $-\cos 2\alpha$, and $\cos(2\alpha - 60°) = \cos 2\alpha \cos 60° + \sin 2\alpha \sin 60° = \frac{1}{2}\cos 2\alpha + \frac{\sqrt{3}}{2}\sin 2\alpha$. Now, observe that $\sin 2\alpha =$

$\sqrt{1-\cos^2 2\alpha}$; this follows from the Pythagorean identity and using the fact that $2\alpha \in (0, \pi)$ (so that $\sin 2\alpha$ is positive). Hence our system of equations implies

$$\begin{aligned} x^2 &= 2 + 2\cos 2\alpha \\ 4x^2 &= 2 - \cos 2\alpha - \sqrt{3}\sqrt{1-\cos^2 2\alpha} \end{aligned}$$

Let $y = \cos 2\alpha$. Multiplying the first equation by 4, we can obtain an equation entirely in terms of y:

$$\begin{aligned} 8 + 8y &= 2 - y - \sqrt{3-3y^2} \\ \sqrt{3-3y^2} &= -6 - 9y \end{aligned}$$

Square both sides to obtain a quadratic in y:

$$\begin{aligned} 3 - 3y^2 &= 81y^2 + 108y + 36 \\ 84y^2 + 108y + 33 &= 0 \\ 28y^2 + 36y + 11 &= 0 \end{aligned}$$

Using the quadratic formula, we obtain $y = -\frac{11}{14}$ or $y = -\frac{1}{2}$, so the possible values of $\cos 2\alpha$ are $-\frac{11}{14}$ or $-\frac{1}{2}$. However, notice that $\frac{1}{2}$ is extraneous as it implies $x = 1 \implies PQ = 3$, which is impossible given the configuration of the problem. Hence $\cos 2\alpha = -\frac{11}{14} \implies x^2 = PA^2 = 2 - 2\left(-\frac{11}{14}\right) = \frac{3}{7}$, so $PA = \frac{\sqrt{21}}{7}$ and $PQ = 3x = \boxed{\frac{3\sqrt{21}}{7}}$. □

21. If $x = y$, then we obtain $f(2x) = 4x^2 + f(0) = 4x^2$ for all integers x. Then $f(x) = x^2$ if x is even.

    If $y = x - 1$, then we obtain $f(2x-1) = 4x(x-1) + f(1) = 4x^2 - 4x + 3 = (2x-1)^2 + 2$ for all integers x. In other words, we can say that $f(x) = x^2 + 2$ if x is odd.

    Therefore, f is defined as follows:

$$f(x) = \begin{cases} x^2 & x \text{ is even} \\ x^2 + 2 & x \text{ is odd} \end{cases}$$

    Then $f(19) + f(20) = (19^2 + 2) + 20^2 = \boxed{763}$. □

22. Consider the set of residues modulo 1000 which are relatively prime to 1000 (namely, $S = \{1, 3, 7, 9, 11, 13, 17, 19, \ldots, 999\}$). We use the fact that for each $x \in S$, there exists exactly one element $x^{-1} \in S$ such that $xx^{-1} \equiv 1 \pmod{1000}$. Specifically, S forms a group under the multiplication operator (modulo 1000), often denoted $(\mathbb{Z}/1000\mathbb{Z})^\times$.

    Observe that if x has units digit 3, then x^{-1} has units digit 7; conversely, if x has units digit 7, then x^{-1} has units digit 3. Hence we can partition S into $\frac{\varphi(1000)}{2} = 200$ pairs (x, y) where x has units digit 3 and $y = x^{-1}$ has units digit 7. The product of each of these pairs is $1 \pmod{1000}$, so the overall product is $1 \pmod{1000}$, so the remainder when N is divided by 1000 is $\boxed{1}$. □

23. We have $P(z) = 1 + z^4 + z^8 + z^{12} = \frac{z^{16}-1}{z^4-1}$. Hence, if $P(z) = 0$, then z is a 16th root of unity, but not a 4th root of unity. Specifically, $z = \text{cis}\left(\frac{k\pi}{8}\right)$ where $k = 1$, 2, 3, 5, 6, 7, 9, 10, 11, 13, 14, 15.

When plotted in the convex plane, the region R will appear as follows, where each vertex has absolute value (norm) equal to 1.

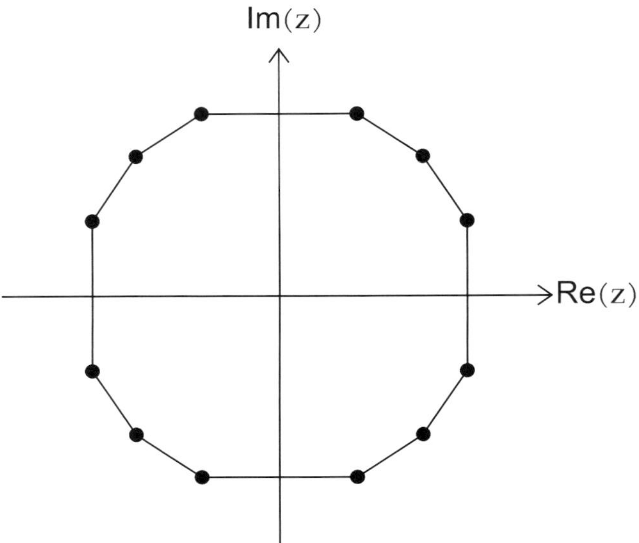

To find the area of R quickly, we subdivide the dodecagon into 12 isosceles triangles as shown:

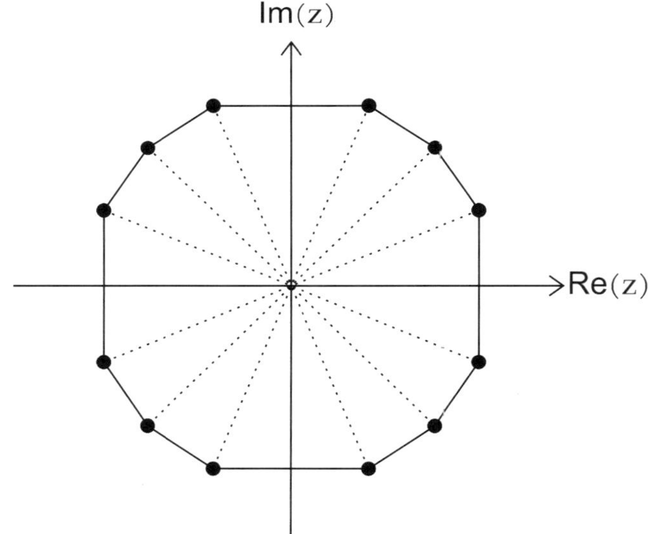

Four of these triangles have sides 1, 1, and vertex angle of $\frac{\pi}{4}$, while the remaining eight have sides 1, 1, and vertex angle $\frac{\pi}{8}$. Using the area formula $\frac{1}{2}ab\sin C$, the area of the dodecagon is $4\left(\frac{1}{2}\sin\frac{\pi}{4}\right)+8\left(\frac{1}{2}\sin\frac{\pi}{8}\right) = 2\sin\frac{\pi}{4}+4\sin\frac{\pi}{8}$. We have $\sin\frac{\pi}{4} = \frac{\sqrt{2}}{2}$ and $\sin\frac{\pi}{8} = \frac{\sqrt{2-\sqrt{2}}}{2}$ using the half angle formula. Substituting these in, we obtain $[R] = \sqrt{2}+2\sqrt{2-\sqrt{2}}$.

Using the approximation $1.41 < \sqrt{2} < 1.42$, we see that $\sqrt{2-\sqrt{2}} < \sqrt{0.59} < 0.77$. Hence $\sqrt{2}+2\sqrt{2-\sqrt{2}} < 1.42+2(0.77) = 2.96$, so the answer choice closest to the actual area of R is $\boxed{2.9}$. □

24. The 3rd coin flip is equally likely to be heads or tails, since $H_2 = 1$ and $\frac{H_2}{3-1} = \frac{1}{2}$.

Given the first n coin flips, we can construct the probability distribution representing the probability of obtaining 0, 1, ..., n heads, which can be represented as an $(n+1)$-dimensional vector. For $n = 3$, this probability distribution is $\left(0, \frac{1}{2}, \frac{1}{2}, 0\right)$. Note that the first and last entries are necessarily zero, since the first two flips are always head followed by tail.

Suppose $n = 4$. The probability that exactly 1 head is flipped equals $\frac{1}{2}$ times the probability that the fourth flip is a tail (conditioned that 1/3 previous flips was heads). This probability is $\frac{1}{2} \times \left(1-\frac{1}{3}\right) = \frac{1}{3}$. Using similar reasoning, the probability that 2/4 heads are flipped is $\frac{1}{2}\cdot\frac{1}{3}+\frac{1}{2}\cdot\left(1-\frac{2}{3}\right) = \frac{1}{3}$. Then the probability that 3/4 heads are flipped is $\frac{1}{3}$, and the probability distribution for $n = 4$ is $\left(0, \frac{1}{3}, \frac{1}{3}, \frac{1}{3}, 0\right)$.

Observe that the probability distributions in both cases are uniform among 1, 2, ..., $n-1$ heads. Indeed, this is the case:

Claim: For $n \geq 3$, the corresponding probability distribution is $\left(0, \frac{1}{n-1}, \frac{1}{n-1}, \dots, \frac{1}{n-1}, 0\right)$.

Proof. By induction on n. Specifically, we show that for all $1 \leq i \leq n$, the probability of obtaining i heads is $\frac{1}{n-1}$.

The statement holds for $n = 2$ and $n = 3$. Suppose true for some $n \geq 3$. We will show that, for all $1 \leq i \leq n+1$, the probability of obtaining i heads out of $n+1$ flips is $\frac{1}{n}$.

To find the probability of obtaining i out of $n+1$ heads, there are two cases: either $i-1$ out of the first n flips are heads and the $(n+1)^{\text{th}}$ flip is head, or i out of the first n flips are heads and the $(n+1)^{\text{th}}$ flip is tail. Using the induction hypothesis, the first case occurs with probability $\frac{1}{n-1}\cdot\frac{i-1}{n}$. The second case occurs with probability $\frac{1}{n-1}\cdot\left(1-\frac{i}{n}\right)$. Adding the probabilities from both cases, the probability of obtaining i heads out of $n+1$ flips is

$$\frac{1}{n-1}\cdot\frac{i-1}{n}+\frac{1}{n-1}\cdot\left(1-\frac{i}{n}\right) = \frac{i-1+n-i}{(n-1)n} = \frac{1}{n}$$

completing the induction step. □

Therefore, with 2020 coin flips, the probability of obtaining 1010 heads is $\frac{1}{2020-1} = \boxed{\frac{1}{2019}}$. □

25. First, we see that $MN = OP = QR = 7\sqrt{7}$; this follows from the law of cosines on $\triangle MBN$ (alternatively, one can drop an altitude from N onto MB and use the Pythagorean theorem).

Extend $\overline{MN}$, $\overline{OP}$, and $\overline{QR}$ to form triangle XYZ. Then ω is the incircle of this triangle. Consider triangle RMX in the figure below:

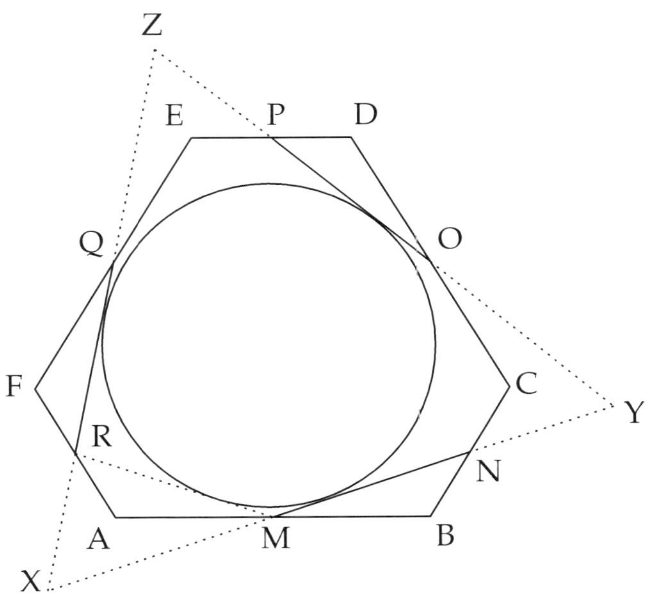

Further, let $\angle RMA = \alpha$ and $\angle MRA = \beta$, where $\alpha + \beta = 120^\circ$. Because $\triangle RMA \cong \triangle NMB$ by SSS congruence, we have $\angle AMX = \angle NMB = \alpha$, and similarly $\angle ARX = \beta$, so it follows that A is the incenter of $\triangle RMX$ as it is the intersection of its angle bisectors. Further, $\angle RXM = 180^\circ - 2\alpha - 2\beta = 60^\circ$; using this fact, we establish that $\triangle XYZ$ is equilateral.

Note that $RM = 7\sqrt{7}$. Using some more congruence, the side length of the desired equilateral triangle is $XR + XM + 7\sqrt{7}$ (specifically, suppose Y is the intersection of the extensions of RQ and OP. Then $\triangle RMX \cong \triangle PQY$, so $QY = MX$, and $XY = XR + RQ + QY = XR + XM + 7\sqrt{7}$).

We first compute the inradius of $\triangle RMX$. Fortunately this is easy to find, as the inradius is simply the altitude from A to RM; denote by H the foot of this altitude. By considering the area of $\triangle RMA$, we have $[\triangle RMA] = \frac{1}{2} \cdot 7 \cdot 14 \cdot \sin 120^\circ = \frac{1}{2} \cdot 7\sqrt{7} \cdot AH$, so $AH = \sqrt{21}$. Using the Pythagorean theorem on $\triangle AHR$ and $\triangle AHM$, we deduce $HR = 2\sqrt{7}$ and $HM = 5\sqrt{7}$.

Let J and K be the feet of the altitudes from A to MX and RX respectively. Then $JM = 5\sqrt{7}$, $JX = XK = 3\sqrt{7}$ (using the fact that $\triangle AJX$ is 30-60-90), and $KR = 2\sqrt{7}$, so we have $XM = 8\sqrt{7}$ and $XR = 5\sqrt{7}$.

Hence the side length of the desired equilateral triangle is $5\sqrt{7} + 8\sqrt{7} + 7\sqrt{7}$, which simplifies to $20\sqrt{7}$. The inradius of an equilateral triangle with side length $20\sqrt{7}$ is $\frac{10\sqrt{21}}{3}$, so the area of ω is $\left(\frac{10\sqrt{21}}{3}\right)^2 \pi = \boxed{\frac{700\pi}{3}}$. □

1. Since
$$\frac{2}{1300(x-2)} = \frac{2}{1300x - 2600} = \frac{1}{650} - \frac{1}{651} = \frac{1}{650 \times 651},$$
we have $\frac{1}{x-2} = \frac{1}{651}$ by simplifying both sides. Thus $x = \boxed{653}$. □

2. Consider the addition of $n+1$. If no carrying is performed, then $s(n+1) = 100$. If carrying is performed, then each carry reduces the sum of the digits by 9. Hence 91 is a possibility for $s(n+1)$. However, it is possible that two or more carries are performed, if the last two (or more) digits of n are 9. So the possible values of $s(n+1)$ are 100, 91, 82, 73, ..., 1, in which the sum is $100 + 91 + 82 + \ldots + 1 = \frac{101 \times 12}{2} = \boxed{606}$. □

3. There are $6 \times 8 = 48$ equally likely outcomes. Doing casework on Nisha's roll, we see that the probability is $\frac{6+6+5+4+4+3+2+1}{48} = \frac{27}{48} = \boxed{\frac{9}{16}}$. □

4. Note that $m\angle BAC = m\angle BCA = 65°$. Also, since $m\angle ADC = 130°$, we have $m\angle CAD = m\angle ACD = 25°$. Therefore $m\angle BAD = 65° - 25° = \boxed{40°}$. □

5. We know the median and unique mode are 90; since the mean is also 90, the sum of the students' scores is $90 \times 7 = 630$.

   In order to minimize the score of one of the students, we greedily maximize everyone else's scores. The following sequence of scores is optimal:
$$61 \quad 90 \quad 90 \quad 90 \quad 99 \quad 100 \quad 100$$
   Note that we cannot have 60, 90, 90, 90, 100, 100, 100 as the scores since the mode is unique. Hence the lowest possible score is $\boxed{61}$. □

6. We have $2^{5n+1} + 5^{n+k} = 2 \times 32^n + 5^{n+k} \equiv 2 \times 5^n + 5^{n+k} \equiv 5^n(2 + 5^k) \pmod{27}$. Hence $2^{5n+1} + 5^{n+k}$ is divisible by 27 if and only if $5^n(2 + 5^k)$ is divisible by 27. As 5^n and 3 are relatively prime, $2 + 5^k$ must be divisible by 27. The only answer choice that satisfies this condition is $k = \boxed{2}$. □

7. Observe that at all times, both ants are either at opposite vertices of the cube, or are on adjacent vertices of the cube (i.e. 1 unit apart). The ants can never be on the same vertex at the same time, so the probability is $\boxed{0}$. □

8. We have the following diagram:

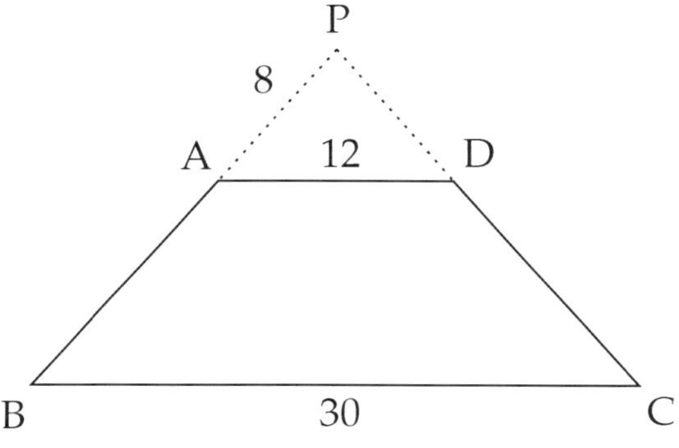

By AAA similarity, we have $\triangle PAD \sim \triangle PBC$, and the ratio of the side lengths is $12 : 30 = 2 : 5$. Letting $AB = x$, we have $\frac{8}{8+x} = \frac{2}{5} \implies x = AB = \boxed{12}$. Since $AB = DC$, we also have $DC = 12$. The perimeter is $30 + 12 + 12 + 12 = \boxed{66}$. □

9. Using the change of base formula, we can rewrite the given equation as

$$\frac{\log x}{\log 2} + \frac{\log x}{\log 4} + \frac{\log x}{\log 8} = \frac{\log x}{\log 2}\frac{\log x}{\log 4}\frac{\log x}{\log 8}$$

where all logs can be taken using base 2. We notice that $\log x = 0 \iff x = 1$ is a solution; if $x \neq 1$, then dividing both sides by $\log x$ gives

$$\frac{1}{\log 2} + \frac{1}{\log 4} + \frac{1}{\log 8} = \frac{(\log x)^2}{\log 2 \log 4 \log 8}$$

or equivalently,

$$\log_2 \log 4 + \log 4 \log 8 + \log 2 \log 8 = (\log x)^2$$

Treating these logarithms in base 2, this gives $2 + 6 + 3 = 11 = (\log_2 x)^2 \iff \log_2 x = \pm\sqrt{11}$. The largest real solution is $x = 2^{\sqrt{11}}$ which is between 2^3 and 2^4, so the answer is $\boxed{[8, 16)}$. □

10. Listing out the terms mod 16, we obtain 1, 1, 4, 10, 12, 12, 0, 8, 0, 0, ...in which all successive terms are $0 \pmod{16}$. The sum is congruent to $1+1+4+10+12+12+0+8 \equiv 48 \equiv \boxed{0} \pmod{16}$. □

11. The coefficient of x^6 represents the number of non-negative integer solutions to the equation $x_1 + x_2 + x_3 + x_4 = 6$ where $0 \leq x_i \leq 3$. We will do some casework:

- **Case 1:** Two of the x_i's are 3. Then the variables are $(3, 3, 0, 0)$ (up to ordering), giving $\binom{4}{2} = 6$ solutions.
- **Case 2:** One of the x_i's equals 3. The variables can be (up to ordering) $(3, 1, 1, 1)$, $(3, 2, 1, 0)$, giving $4 + 24 = 28$ solutions.
- **Case 3:** None of the x_i's are equal to 3. The variables are (up to ordering) $(2, 2, 1, 1)$ or $(2, 2, 2, 0)$, giving $6 + 4 = 10$ solutions.

The number of integer solutions, and hence the x^6 coefficient, is $6 + 28 + 10 = \boxed{44}$.

Alternate solution: Consider the generating function $G(x) = (1 + x + x^2 + x^3)^4$, which can be written as $\left(\frac{1-x^4}{1-x}\right)^4 = (1-x^4)^4\frac{1}{(1-x)^4}$. We may substitute $\frac{1}{(1-x)^4}$ with $\binom{3}{3} + \binom{4}{3}x + \binom{5}{3}x^2 + \binom{6}{3}x^3 + \ldots$, so that we equivalently seek the x^6 coefficient of the expression

$$(1 - x^4)^4\left(\binom{3}{3} + \binom{4}{3}x + \binom{5}{3}x^2 + \binom{6}{3}x^3 + \ldots\right).$$

Since $(1-x^4)^4 = 1-4x^4+6x^8-4x^{12}+x^{16}$, we obtain that the x^6 coefficient is $\binom{9}{3} - 4\binom{5}{3} = 84 - 40 = \boxed{44}$. □

12. Since $AK = AM$, we get $AK = 8$, $KB = BL = 3$, and $LC = LM = 7$. Hence the side lengths of $\triangle ABC$ are 10, 11, and 15. Using Heron's formula with $s = 18$, the area of $\triangle ABC$ is

$$[\triangle ABC] = \sqrt{18(8)(7)(3)} = \boxed{12\sqrt{21}}.$$

□

13. We can specify $f(f(n))$ unambiguously depending on the value of $n \pmod 4$:

$$f(f(n)) = \begin{cases} \frac{n}{4} & n \equiv 0 \pmod 4 \\ \frac{n}{2}+1 & n \equiv 2 \pmod 4 \\ \frac{n+1}{2} & n \text{ is odd} \end{cases}$$

Hence we have

$$\begin{aligned}\sum_{n=1}^{2020} f(f(n)) &= [f(f(4)) + f(f(8)) + \ldots + f(f(2020))] \\ &+ [f(f(2)) + f(f(6)) + \ldots + f(f(2018))] \\ &+ [f(f(1)) + f(f(3)) + \ldots + f(f(2019))] \\ &= (1+2+3+\ldots+505) \\ &+ (2+4+6+\ldots+1010) \\ &+ (1+2+3+\ldots+1010) \\ &= 3\sum_{i=1}^{505} i + \sum_{i=1}^{1010} i \\ &= \frac{3 \times 505 \times 506}{2} + \frac{1010 \times 1011}{2} \\ &= \frac{505(1518+2022)}{2} = 505 \times 1770 = \boxed{893850}.\end{aligned}$$

□

14. We observe the differences between the prime factorizations of $20!$ and $21!$. Namely, they differ by one power of 3 and one power of 7: $20! = 3^8 \times 7^2 \times N$ and $21! = 3^9 \times 7^3 \times N$ where N is not divisible by 3 or 7. Then in order for lcm$(20!, n)$ to equal $21!$, n must contain 3^9 and 7^3 in its prime factorization. The smallest such n is $3^9 \times 7^3$, and the number of divisors of n is $(9+1)(3+1) = \boxed{40}$. □

15. There are 2^8 equally likely outcomes corresponding to the possible sequences of true/false the student can answer. Since question n is worth n points, we can put the true/false sequences in 1-to-1 correspondence with subsets S of $\{1,2,3,\ldots,8\}$ where $n \in S$ iff question n was answered correctly.

Hence we can count the number of subsets of $\{1,2,3,\ldots,8\}$ whose sum is at least 18. Since $1+2+3+\ldots+8 = 36$, the number of subsets is approximately half the total number of subsets. We observe that the number of subsets with sum ≥ 19 is equal to the number of subsets with sum ≤ 17, so it suffices to count the number of subsets S of $\{1,2,3,\ldots,8\}$ with sum equal to 18.

To do this, we can use casework on the largest element of S. If the largest element is 8, then we have 7 subsets: $\{3,7,8\}$, $\{1,2,7,8\}$, $\{4,6,8\}$, $\{1,3,6,8\}$, $\{1,4,5,8\}$, $\{2,3,5,8\}$, $\{1,2,3,4,8\}$. If the largest element is 7, then we have 5 subsets: $\{5,6,7\}$, $\{1,4,6,7\}$, $\{2,3,6,7\}$, $\{2,4,5,7\}$, $\{1,2,3,5,7\}$. If the largest element is 6, we have 2 subsets: $\{1,2,4,5,6\}$ and $\{3,4,5,6\}$. The largest element of S cannot be 5 or less since $1+2+3+4+5 = 15 < 18$. Altogether there are $7 + 5 + 2 = 14$ subsets of $\{1,2,3,\ldots,8\}$ with sum equal to 18.

Out of the remaining $2^8 - 14 = 242$ remaining subsets, exactly half (or 121) have sum strictly greater than 18. Then there are $14 + 121 = 135$ subsets of $\{1,2,3,\ldots,8\}$ whose sum is at least 18, and the probability is $\boxed{\frac{135}{256}}$. □

16. Let O_1 and O_2 denote the centers of ω_1 and ω_2, and let P and Q be the intersection points so that A, P, Q, B are in that order.

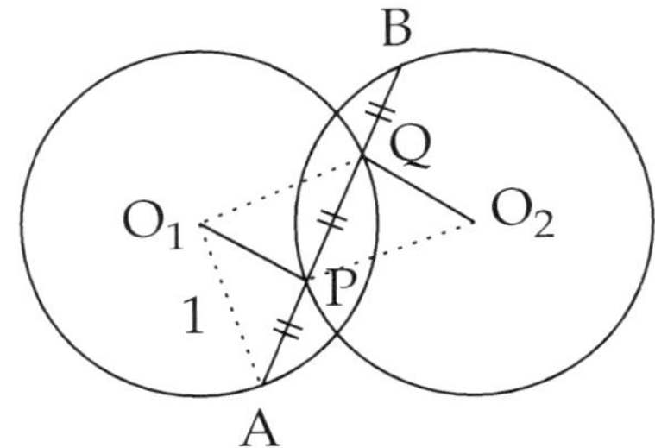

Since P is the midpoint of chord AQ (of circle ω_1, it follows that $\angle O_1PA = 90°$. Further, $O_1Q = O_2P = 1$ and $O_1P = O_2Q$, so O_1PO_2Q is a parallelogram.

Consider parallelogram O_1PO_2Q. Recall $O_1O_2 = \frac{3}{2}$ and $\angle O_1PQ = \angle O_2QP = 90°$.

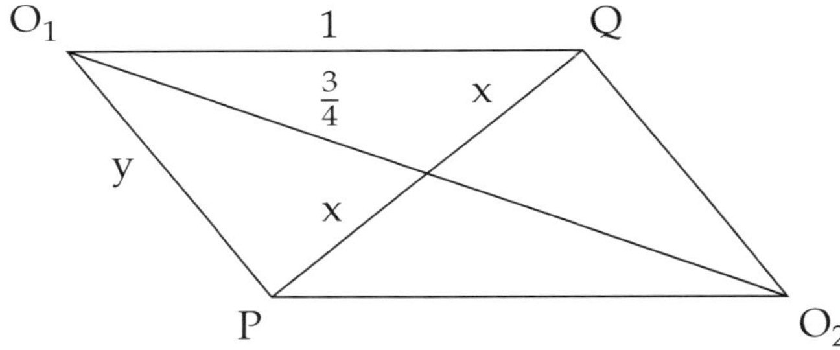

Letting $PQ = 2x$ and $O_1P = y$, we have by the Pythagorean theorem $y^2 + x^2 = \frac{9}{16}$ and $y^2 + (2x)^2 = 1$. Subtracting the first equation from the second yields $3x^2 = \frac{7}{16}$, or $x = \frac{\sqrt{21}}{12}$.

Then $PQ = 2x = \frac{\sqrt{21}}{6}$, and $AB = 3PQ = \boxed{\frac{\sqrt{21}}{2}}$. □

17. Suppose the "base" of this triangle is the side with endpoints $(0,0)$ and $(2,4)$, and length $2\sqrt{5}$. The altitude h of this triangle is the distance from the point (x, x^2) to the line $y = 2x$. Using the formula for the distance from a point to a line:

$$h = \frac{|-2x + x^2|}{\sqrt{(-2)^2 + 1^2}} = \frac{|-2x + x^2|}{\sqrt{5}}$$

Note that, on the interval $(0, 2)$, the value of $-2x + x^2$ is negative, so $|-2x + x^2| = 2x - x^2$. The value of x which maximizes h is $x = 1$, yielding $h = \frac{1}{\sqrt{5}}$. The maximum area is $\frac{1}{2} \cdot 2\sqrt{5} \cdot \frac{1}{\sqrt{5}} = \boxed{1}$. □

18. We observe that $k = 1$ is a trivial solution. By Euler's totient theorem, $11^{40} \equiv 1 \pmod{100}$ and $13^{40} \equiv 1 \pmod{100}$, so $11^{40} + 13^{40} \equiv 2 \pmod{100}$. More generally, $11^{40m} + 13^{40m} \equiv 2 \pmod{100}$ and $11^{40m+1} + 13^{40m+1} \equiv 11 + 13 \equiv 24 \pmod{100}$.

    Since 81 is of the form $40m + 1$ for some integer m, we see that $k = \boxed{81}$ works. Further, it can be checked that none of the other choices work. □

19. Label the children A, B, ..., E where child A has 5 candies, child B has 10 candies, ..., child E has 25 candies. Note that $\frac{5+10+15+20+25}{5} = 15$. We observe that if it is possible for the children to distribute candies to each other such that they have an equal number of candies, then at least 10 moves are required; this is because child A initially has 5 candies and can gain at most one candy on each move.

    We show that 10 moves is indeed possible. Let a denote the number of moves for which child A gives one candy to each of the remaining children. Define b, c, d, and e similarly, and let $M = a + b + c + d + e$ denote the number of moves until they each have the same number of candies. We see that child A will have $5 + (b + c + d + e - 4a)$ candies, which may be written alternatively as $5 + (M - 5a)$. Thus we want to find (a, b, c, d, e) such that the following equations hold:

$$\begin{aligned} 5 + (M - 5a) &= 15 \\ 10 + (M - 5b) &= 15 \\ 15 + (M - 5c) &= 15 \\ 20 + (M - 5d) &= 15 \\ 25 + (M - 5e) &= 15 \end{aligned}$$

    The third equation implies $M - 5c = 0$, so M is a multiple of 5. Consider $(a, b, c, d, e) = (0, 1, 2, 3, 4)$ and $M = 10$. It can be checked that any valid sequence of moves where A gives candies 0 times, B gives candies once, ..., E gives candies four times results in the situation where the children all have the same number of candies. Hence the minimum number of moves is $\boxed{10}$. □

20. Let $\angle HCD = \angle DCA = \alpha$, $CH = h$, and $AD = x$. Because $\triangle ACH \sim \triangle ABC$, we have $\angle B = 2\alpha$. Also, by the angle bisector theorem, $AC = hx$.

    Note that $\tan \alpha = \frac{1}{h}$. We can express $\tan 2\alpha$ two different ways:

$$\tan 2\alpha = \frac{hx}{5} = \frac{\frac{2}{h}}{1 - \frac{1}{h^2}} = \frac{2h}{h^2 - 1}$$

    Cross-multiplying yields $10h = hx(h^2 - 1)$ or $x(h^2 - 1) = 10$.

    Using the similarity of triangles CBH and ACH, we also establish

$$\frac{h}{5} = \frac{x+1}{hx}$$

Cross-multiplying yields $h^2x = 5x + 5$, or $h^2 = 5 + \frac{5}{x}$. Now we have h^2 in terms of x; substituting into our previous equation yields

$$x\left(5 + \frac{5}{x} - 1\right) = 10$$

which simplifies to $5 + 4x = 10 \implies x = \frac{5}{4}$. Then $h = 3$, and $AD + AC = x + hx = \frac{5}{4} + \frac{15}{4} = \boxed{5}$. □

21. The probability that David initially rolls a sum of 11 or greater is $\frac{1}{2}$ (this can be seen by symmetry, since the probability of rolling 11, 12, ..., 18 is the same as the probability of rolling 10, 9, ..., 3).

Otherwise, David rolls 10 or less, in which he rerolls the lowest numbered die. We will do casework on the sum of the highest two dice rolls. Observe that if David's highest two dice rolls sum to ≥ 10, then he already wins. If David's highest two dice rolls sum to 9, then he wins with probability $\frac{5}{6}$ (by rolling 2 or greater). More generally, if David's highest two dice rolls sum to i, for $i = 5, \dots, 9$, then he wins with probability $\frac{i-4}{6}$. We perform the casework:

- **Case 1: The sum of David's highest two dice rolls is 9.** The possible dice rolls (up to ordering) are $(?, 3, 6)$ and $(?, 4, 5)$ where ? represents the die that is to be rerolled. Out of $6^3 = 216$ outcomes, there are $(6 + 6 + 3) + (6 + 6 + 6 + 3) = 36$ outcomes satisfying this case.
- **Case 2: The sum of David's highest two dice rolls is 8.** The possible dice rolls are $(?, 2, 6)$, $(?, 3, 5)$, $(?, 4, 4)$, giving $(6 + 3) + (6 + 6 + 3) + (3 + 3 + 3 + 1) = 34$ outcomes.
- **Case 3: The sum of David's highest two dice rolls is 7.** The possible dice rolls are $(?, 1, 6)$, $(?, 2, 5)$, $(?, 3, 4)$, giving $3 + (6 + 3) + (6 + 6 + 3) = 27$ outcomes.
- **Case 4: The sum of David's highest two dice rolls is 6.** The possible dice rolls are $(?, 1, 5)$, $(?, 2, 4)$, $(?, 3, 3)$, giving $3 + (6 + 3) + (3 + 3 + 1) = 19$ outcomes.
- **Case 5: The sum of David's highest two dice rolls is 5.** The possible dice rolls are $(?, 1, 4)$, $(?, 2, 3)$, giving $3 + (6 + 3) = 12$ outcomes.

The desired probability is

$$\frac{1}{2} + \frac{36 \cdot 5 + 34 \cdot 4 + 27 \cdot 3 + 19 \cdot 2 + 12 \cdot 1}{6^4} = \frac{365}{432}$$

so $m + n = 365 + 432 = \boxed{797}$. □

22. Let $S = \sum_{i=1}^{\infty} \frac{a_i}{3^i} = \frac{1}{3} + \frac{1}{9} + \frac{3}{27} + \frac{7}{81} + \frac{17}{243} + \dots$. It can be shown that this sum converges, as the characteristic polynomial corresponding to this recursion is $x^2 - 2x - 1$ whose roots are $1 \pm \sqrt{2}$. This implies that the general formula for a_n is of the form $a(1 + \sqrt{2})^n + b(1 - \sqrt{2})^n$ for some constants a and b, so the sequence $\left(\frac{a_n}{3^n}\right)$ is the sum of two geometric sequences with common ratio whose absolute value is less than 1.

We will multiply both sides by 6:

$$\begin{aligned} S &= \frac{1}{3} + \frac{1}{9} + \frac{3}{27} + \frac{7}{81} + \frac{17}{243} + \ldots \\ 6S &= 2 + \frac{2}{3} + \frac{6}{9} + \frac{14}{27} + \frac{34}{81} + \ldots \end{aligned}$$

Adding these two equations:

$$\begin{aligned} 7S &= 2 + \left(\frac{3}{3} + \frac{7}{9} + \frac{17}{27} + \frac{41}{81} + \ldots\right) \\ &= 2 + (9S - 4) \end{aligned}$$

Hence $7S = 9S - 2$, or $S = \boxed{1}$. □

23. We will find the minimum possible distance between a point on either parabola to the origin, as this determines the maximum possible radius of the circle. Note that, by symmetry about the x- and y-axes, we can simply find the minimum possible distance between the origin and the parabola $y = x^2 - 4$ on the interval $x \in [0, 2]$.

The square of the distance between the origin and the point $(x, x^2 - 4)$ is given by

$$\begin{aligned} d^2 &= x^2 + (x^2 - 4)^2 \\ &= x^4 - 7x^2 + 16 \end{aligned}$$

Note that for real z, the minimum value of $z^2 - 7z + 16$ occurs at $z = \frac{7}{2}$. In this case, setting $x^2 = \frac{7}{2}$ achieves the global minimum of $d^2 = \frac{15}{4}$. The area of this circle is $d^2\pi = \boxed{\frac{15}{4}\pi}$. □

24. We use the fact that AJ_A bisects $\angle A$. This property is true of all excircles; let P and Q are the points where the A-excircle is tangent to the extensions of AB and AC respectively. Then $\triangle APJ_A \cong \triangle AQJ_A$.

Further, since $CQ + BP = 14$ and $15 + CQ = 13 + BP$, we establish that $BP = 8$ and $CQ = 6$ as shown in the following diagram:

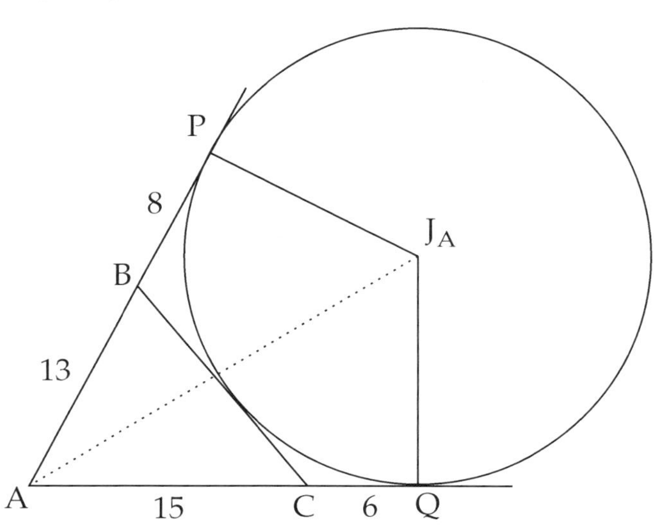

Let D be the foot of the altitude from B to AC, and let E be the intersection of BD and AC. Then $\triangle ADE \sim \triangle AQJ_A$. Since the area of a 13-14-15 triangle is 84, we have $BD = \frac{56}{5}$, so by the Pythagorean theorem, $AD = \frac{33}{5}$.

Then by the angle bisector theorem (as AJ_A bisects $\angle A$), we have $DE : EB = 33 : 65$; since $BD = \frac{56}{5}$, we establish $DE = \frac{33}{98} \times \frac{56}{5} = \frac{132}{35}$. By the Pythagorean theorem on $\triangle ADE$, $AE = \frac{33\sqrt{65}}{35}$ (tedious calculation can be avoided by noticing that the side lengths are a 4:7:$\sqrt{65}$ right triangle, scaled by $\frac{33}{35}$).

In particular, $\cos \angle DAE = \cos \angle QAJ_A = \frac{7\sqrt{65}}{65}$. Hence we have $\frac{7}{\sqrt{65}} = \frac{21}{AJ_A}$, so $AJ_A = \boxed{3\sqrt{65}}$.

Alternate solution: Using the formula for the radius of the A-excircle, we have

$$R_A = \frac{rs}{s-a} = \frac{84}{21-14} = 12$$

Since $CQ = 6$, then $AQ = 21$, so by the Pythagorean theorem, $AJ_A = \sqrt{21^2 + 12^2} = \boxed{3\sqrt{65}}$. $\square$

25. We will consider the problem in binary. The binary representation of 2020 is 11111100100_2.

Let $f(n)$ denote the number of tuples of the form $(a_0, a_1, \dots, a_k)$ such that $\sum_{i=0}^{k} a_i 2^i = n$ and $a_i \in \{0, 1, 2\}$. The desired answer is $f(2020)$ (note that $2^{11} > 2020$, so the maximum possible k is 10). In order to express the tuples concisely, we will invent a "pseudobinary" which is exactly like binary, except that the digits may be 0, 1, or 2 (e.g., 201 corresponds to $2 \times 2^2 + 0 \times 2^1 + 1 = 9$, and the pseudobinary representation may not be unique).

n	$f(n)$	Pseudobinary	n	$f(n)$	Pseudobinary
1	1	1	9	3	1001, 201, 121
2	2	2, 10	10	5	1010, 1002, 210, 202, 122
3	1	11	11	2	1011, 211
4	3	100, 12, 20	12	5	1100, 1020, 220, 1012, 212
5	2	101, 21	13	3	1101, 1021, 221
6	3	110, 102, 22	14	4	1110, 1102, 1022, 222
7	1	111	15	1	1111
8	4	1000, 200, 120, 112	16	5	10000, 2000, 1200, 1120, 1112

We observe some patterns regarding f. First, we see that $f(2^k - 1) = 1$ (the only "pseudobinary" representation of $2^k - 1$ is 11...1), and $f(2^k) = k + 1$. We can state these claims more generally:

Claim: The following are true for any integer n:

$$f(2n) = f(n) + f(n-1) \quad (1)$$
$$f(2n+1) = f(n) \quad (2)$$

To see why these claims are true, note that the pseudobinary representation of $2n$ can be obtained by appending a 0 to any representation of n, or by appending a 2 to any

representation of $n-1$ (which proves (1)), and a pseudobinary representation of $2n+1$ must be obtained by appending a 1 to any representation of n.

Corollary: $f(2^k-2)=k$.

This can be shown inductively with (1).

Corollary: $f(4n)=f(n)+2f(n-1)$.

This can be shown by applying (1) then (2).

Note that $2020 = 11111100100_2$. Using the above claim and corollaries, we compute $f(2020)$ successively:

$$\begin{aligned}
f(63) &= f(111111_2) = 1 \\
f(252) &= f(11111100_2) = f(63) + 2f(62) \\
&= 1 + 12 = 13 \\
f(505) &= f(111111001_2) = f(252) = 13 \\
f(2020) &= f(11111100100_2) = f(505) + 2f(504) \\
&= 13 + 2(f(252) + f(251)) \\
&= 13 + 2(13 + f(62)) && \text{(using (2), } f(251) = f(125) = f(62)\text{)} \\
&= 13 + 2(13 + 6) = \boxed{51}
\end{aligned}$$

□

Solutions for Practice Exam 10.

1. $\frac{3^{20}-3^{10}}{(3^5-1)(3^5+1)} = \frac{3^{10}(3^{10}-1)}{3^{10}-1} = \boxed{3^{10}}$. □

2. Observe that 245 is not divisible by 9, but is divisible by 7. Therefore Maria must have added 8 sometime after multiplying by 9 (otherwise Maria would have obtained a multiple of 9). Further, she must have multiplied by 7 after adding 8 (otherwise Maria would not have obtained a multiple of 7). Hence the only possibility is that Maria multiplied by 9, then added 8, then multiplied by 7. Working backwards, we see that Maria was given the number 3.

   The correct answer is $((3 \times 7) + 8) \times 9 = 29 \times 9 = \boxed{261}$. □

3. Since 2 is the only even prime digit, the number must contain either zero 2's, two 2's, or four 2's. We do casework on the number of 2's in the number. Note that the remaining prime digits are 3, 5, 7.

   If the four-digit integer has no 2's, then there are $3^4 = 81$ such 4-digit integers.

   If the four-digit integer has two 2's, then there are 3 ways to choose the first odd digit, 3 ways to choose the second odd digit, then $\binom{4}{2} = 6$ ways to choose the positions of the two 2's, giving $3 \times 3 \times 6 = 54$ integers.

   If the four-digit integer has four 2's, the only possibility is 2222, or 1 integer.

   Adding up the possibilities for the three cases, we get $81 + 54 + 1 = \boxed{136}$ integers. □

4. Let $\angle DCB = \alpha$ and $\angle DCA = \beta$. Because $\triangle ABC$ is isosceles with $AB = AC$, we have $\angle BAC = 180^\circ - 2(\alpha + \beta)$, so $\angle EAC = 2(\alpha + \beta)$. Also, $\angle ADC = 180^\circ - 2\beta$. The sum of these two angles is 230°, so we have

$$2(\alpha + \beta) + (180^\circ - 2\beta) = 230^\circ$$

   Simplifying, we get $2\alpha = 50^\circ$, or $\alpha = \boxed{25^\circ}$. □

5. The children count 1, 2, 3, ...until nine children are eliminated. The numbers which eliminate children from the circle are 5, 7, 10, 14, 15, 20, 21, 25, and 28. Hence $\boxed{28}$ is the last number counted. □

6. We see that a must be 2 or 3, since a is a prime divisor of 18. If $a = 2$, then $c - b = 9$, which yields no solutions as c and b must be odd primes. Hence $a = 3$, and b is either 2 or 5. If $b = 2$, then $c - a = 20 \implies c = 23$, which does not satisfy the first equation. If $b = 5$, then $c - a = 8 \implies c = 11$, which does satisfy the first equation. Then $(a, b, c) = (3, 5, 11)$ and $a + b + c = 3 + 5 + 11 = \boxed{19}$. □

7. Suppose all six tiles are distinguishable (e.g. the A's are labeled A_1 and A_2). There are $\binom{6}{3} = 20$ equally likely outcomes representing the three tiles that Hannah draws. Of these, exactly $2^3 = 8$ contain one A, one H, and one N. The probability is $\frac{8}{20} = \boxed{\frac{2}{5}}$. □

8. Unfold the surface of the prism so that we obtain the following net:

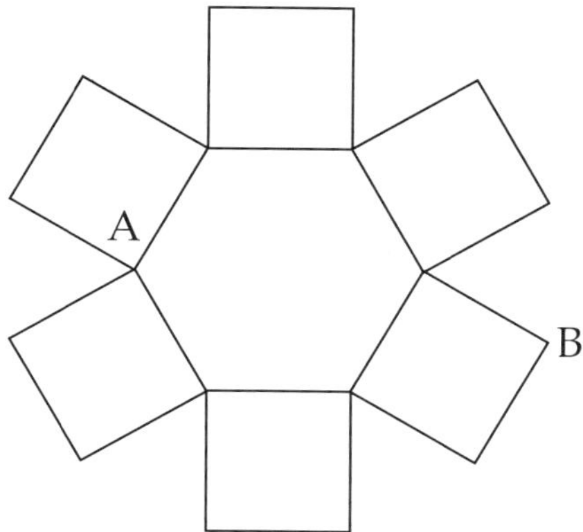

Note that B coincides with two vertices on this net. The minimum distance that the ant crawls equals the distance from A to B on this net. Label points P and Q as shown:

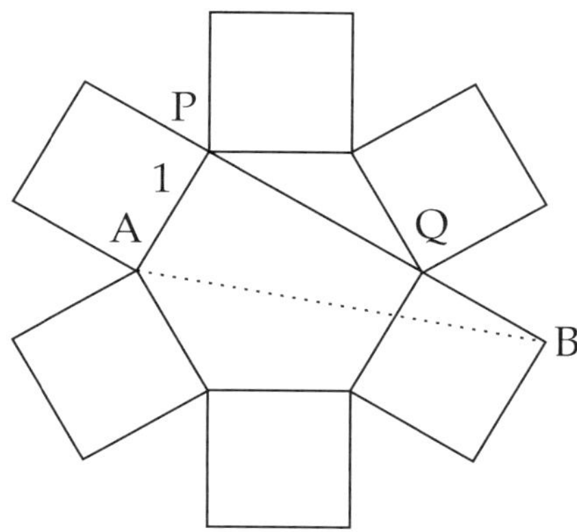

We have $QB = 1$ and $PQ = \sqrt{3}$, so $PB = 1 + \sqrt{3}$. By the Pythagorean theorem, $AB = \sqrt{1^2 + (1+\sqrt{3})^2} = \boxed{\sqrt{5+2\sqrt{3}}}$. □

9. Because the sum of all three roots is -10 (by Vieta's formulas, including multiplicity) which is an integer, the third root must be an integer. Since it is given that the polynomial has exactly two distinct integer roots, one of the roots must have multiplicity 2, say r.

Then by Vieta's formulas, we have

$$\begin{aligned} 2r + s &= -10 \\ r^2 + 2rs &= -32 \\ r^2 s &= 384 \end{aligned}$$

There are many ways to solve for r and s; one way is to substitute $s = -10 - 2r$ into the second equation to obtain a quadratic in r:

$$r^2 + 2r(-10 - 2r) = -32$$

which simplifies to $3r^2 + 20r - 32 = 0$. By the quadratic formula, $r = \frac{-20 \pm \sqrt{400+384}}{6} = -8$ or $\frac{4}{3}$; since it is given that r is an integer, we must have $r = -8$. Then $s = 6$, and $r + s = -8 + 6 = \boxed{-2}$. □

10. We use the following fact:

Claim: For every positive integer k, there exists exactly one power of 2 containing k digits, whose leftmost digit is 1.

Proof. Suppose otherwise, that there exists k for which there is no power of 2 with k digits and leftmost digit 1. Then there exists an exponent a for which $2^a < 10^{k-1} < 2 \times 10^{k-1} \le 2^{a+1}$, which implies $\frac{2^{a+1}}{2^a} > 2$, a contradiction. Hence there must exist a k-digit power of 2 with leftmost digit 1. Further, it can be shown that there must exist exactly one k-digit power of 2. □

Using this fact, we see that there exists exactly one k-digit power of 2 for $k = 2, 3, \ldots, 30$ (we do not count $k = 1$ since $2^0 = 1$ is not counted, and we do not count $k = 31$ since 2^{100} is not counted). There are $30 - 2 + 1 = 29$ such values of k, so the answer is $\boxed{29}$. □

11. The given conditions tell us that the union of all three sets A, B, and C is $\{1, 2, 3, 4, 5\}$, and that any two sets share exactly one element in common (however, it is not necessarily true that A, B, and C share the same element in common).

Suppose that sets A, B, and C satisfy the given conditions, and that $A \cap B = B \cap C = \{i\}$, for some $i \in S$. Then A, B, and C all contain i, so $C \cap A = \{i\}$. If this is the case, there are 5 ways to choose the element common to all three sets. The remaining elements $S \ \{i\}$ must go to exactly one of the three sets A, B, C, so the number of triples of sets in this case is $5 \times 3^4 = 405$.

Otherwise, suppose that the elements contained in $A \cup B$, $B \cup C$, and $C \cup A$ are all different. There are $5 \times 4 \times 3 = 60$ ways to choose which elements belong to the unions of these sets. The remaining two elements must go to exactly one set each, giving $60 \times 3^2 = 540$ triples of sets.

Altogether, we get $405 + 540 = \boxed{945}$ triples of sets. □

12. Let F be the foot of the altitude from D onto BC, as shown:

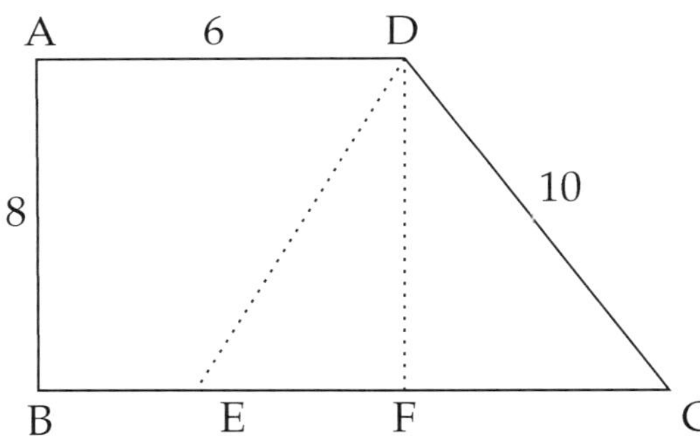

Then $DF = 8$, so $FC = 6$, $EF = 10 - 6 = 4$, and $BE = 6 - 4 = 2$. The area of trapezoid $ABED$ is $\frac{1}{2}(8)(6 + 2) = \boxed{32}$. □

13. Let $\sqrt{19 - 8\sqrt{3}} = a + b\sqrt{3}$. Squaring both sides yields $19 - 8\sqrt{3} = a^2 + 3b^2 + 2ab\sqrt{3}$. Hence we need to find integers a, b with $a > 0$ such that $a^2 + 3b^2 = 19$ and $2ab = -8 \iff ab = -4$. Since a is positive, the only candidates for b are -1, -2, and -4. Testing, we see that $(a, b) = (4, -1)$ works, so $\sqrt{19 - 8\sqrt{3}} = 4 - \sqrt{3}$, and $a + b = 4 + (-1) = \boxed{3}$. □

MATHTOPIA PRESS

14. By Fermat's little theorem, $a^{12} \equiv 1 \pmod{13}$ if a is not divisible by 13, so $a^{2019} = (a^{12})^{168}a^3 \equiv a^3 \pmod{13}$. Then $2^{2019} + 3^{2019} + \equiv +6^{2019} \equiv 2^3 + 3^3 + ... + 6^3 \pmod{13}$.

    To compute $2^3+3^3+...+6^3$ quickly, we can use the sum of cubes formula: $1^3+2^3+...+6^3 = (1+2+...+6)^2 = 21^2 \equiv 12 \pmod{13}$. Subtracting 1, we obtain $2^3 + 3^3 + ... + 6^3 \equiv \boxed{11} \pmod{13}$. □

15. The prime factorization of 2020 is $2^2 \times 5 \times 101$.

    Note that we can think of the problem equivalent to "distributing" two 2's, one 5, and one 101 to the variables x, y, z. There are 3 ways to distribute the 5 and 3 ways to distribute the 101. Using a simple stars and bars argument, there are $\binom{4}{2} = 6$ ways to distribute two 2's to the variables x, y, z. This uniquely determines (x, y, z), so the number of ways is $3 \times 3 \times 6 = \boxed{54}$. □

16. We have the following diagram:

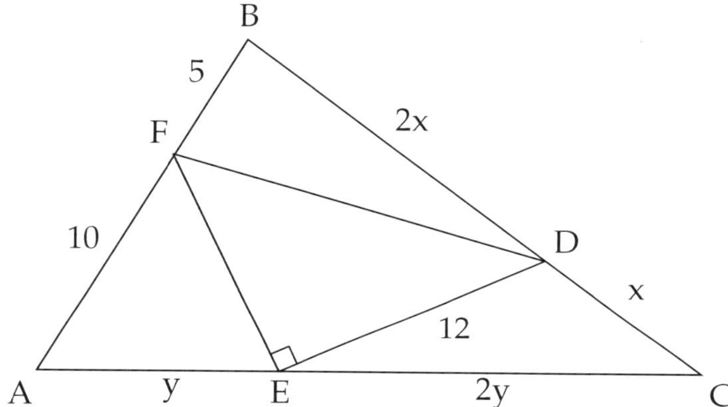

Suppose the area of $\triangle ABC$ is K. We can easily see that $[\triangle BDF] = [\triangle CDE] = [\triangle AEF] = \frac{2}{9}K$ (e.g. using the area formula $\frac{1}{2}ab\sin C$), so $[\triangle DEF] = K - \frac{2}{9}K - \frac{2}{9}K - \frac{2}{9}K = \frac{1}{3}K$. Hence it suffices to find EF, which will allow us to find $[\triangle DEF]$ and $[\triangle ABC]$.

Let P be on $\overline{BD}$ so that $EP \parallel AB$, and let Q be the intersection of $\overline{EP}$ and $\overline{DF}$:

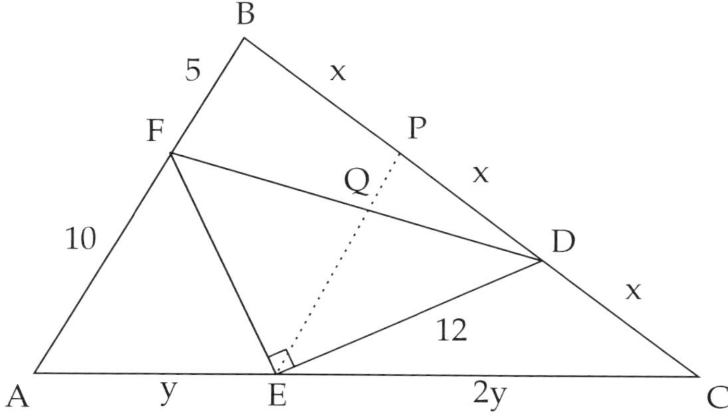

By AAA similarity, $\triangle EPC \sim \triangle ABC$ with $2:3$ ratio, so $BQ = QD = x$, and $EP = 10$. Also by AAA similarity, $\triangle PQD \sim \triangle BFD$ with ratio $1:2$, so $\overline{PQ} = \frac{5}{2}$, and $EQ = 10 - \frac{5}{2} = \frac{15}{2}$.

Further, since Q is the midpoint of hypotenuse DF, we also have $FQ = QD = \frac{15}{2}$. Then $DF = 15$ and $EF = \sqrt{15^2 - 12^2} = 9$, so $[\triangle DEF] = \frac{1}{2} \times 9 \times 12 = 54$. Since $[\triangle DEF] = \frac{1}{3}K$, we have $[\triangle ABC] = 3 \times 54 = \boxed{162}$. ☐

17. Rewrite the first equation as $ab - a - b = 20$. Add 1 to both sides, so that the left-hand side factors to $(a-1)(b-1) = 21$. Repeat this for the remaining equations so that we obtain the equivalent system of three equations:

$$(a-1)(b-1) = 21$$
$$(b-1)(c-1) = 18$$
$$(a-1)(c-1) = 42$$

Multiplying all three equations yields $(a-1)^2(b-1)^2(c-1)^2 = 21 \cdot 18 \cdot 42 \implies (a-1)(b-1)(c-1) = \sqrt{21 \cdot 18 \cdot 42} = 126$. Then $c - 1 = 6$, $a - 1 = 7$, and $b - 1 = 3$ giving $(a,b,c) = (8,4,7)$ and $a + b + c = \boxed{19}$. ☐

18. Note that N is necessarily a 3-digit palindrome; if N is a 2-digit palindrome, then N is a multiple of 11, so $N - 2$ cannot be a multiple of 11. If N is a 1-digit palindrome (i.e. $N = 1, 2, \ldots, 9$), then $N - 2$ also cannot be a multiple of 11 unless $N = 2$, which does not work.

Therefore, let $N = \overline{ABA}$ where A and B are digits and $1 \le A \le 5$. The given conditions tell us that $N \equiv 1 \pmod 9$ and $N \equiv 2 \pmod{11}$; using the divisibility rules for 9 and 11, we see that $2A + B \equiv 1 \pmod 9$ and $2A - B \equiv 2 \pmod{11}$. The only possible value for $2A + B$ is 10 (1 does not work as this implies $A = 0$; 19 does not work unless $N = 595$, which is greater than 500). The only possible value for $2A - B$ is 2 (-9 does not work as this implies $A = 0$; 13 does not work as this implies $A > 5$).

Hence we have $2A + B = 10$ and $2A - B = 2$. Solving the system yields $(A, B) = (3, 4)$, or $N = 343$. The tens digit is $\boxed{4}$. ☐

19. Let $n \le 1000$ be a positive integer. Suppose n is *makeable* if n cents can be made using exactly five coins. We want to determine which values of n are makeable.

Consider the binary representation of n. Immediately, we observe that if the binary representation of n contains five 1's, then n is makeable. However, we make the following stronger observations:

Claim: If $n \ge 5$, and n has at most five 1's in its binary representation, then n is makeable.

Proof idea. The main idea is that a k-cent coin (where $k = 2, 4, 8, \ldots$ can be split into two $(k/2)$-cent coins, which increases the number of coins by 1. Thus, if n has $d \le 5$ 1's in its binary representation, we can replace a coin with two coins of equal value until we have exactly five coins. ☐

Claim: If $n \le 4$ or if n has more than five 1's in its binary representation, then n is not makeable.

Proof idea. Suppose otherwise n has more than five 1's in its binary representation, and n is makeable. We can use a similar idea as in the previous claim but in reverse: replace two $(k/2)$–cent coins with a k-cent coin until we have coins of different denomination. Applying this process, we obtain a binary representation of n containing at most five 1's. However, the binary representation of n contains more than five 1's, a contradiction. □

The above two claims imply that n is makeable if and only if $n \geq 5$ and the binary representation of n contains at most five 1's. Hence we've reduced the problem to counting the number of integers greater than or equal to 5 and less than or equal to 1000 whose binary representation contains at most five 1's. Here, we will treat each integer $n \in \{1, 2, \ldots, 1023\}$ as a length-10 binary sequence where leading zeros are allowed. We observe that $1000 = 1111101000_2$, so any integer in $\{1000, \ldots, 1023\}$ has at least six 1's in binary, and is not counted.

Therefore we freely count $\binom{10}{5}+\binom{10}{4}+\binom{10}{3}+\binom{10}{2}+\binom{10}{1} = 637$ binary sequences containing at least one 1 and at most five 1's. We subtract 4 to account for the numbers 1, ..., 4 (which are not makeable), giving $637 - 4 = \boxed{633}$ monetary amounts. □

20. Because two of the opposite angles ($\angle ABC$ and $\angle ADC$) are right angles, quadrilateral $ABCD$ must be cyclic with diameter $\overline{AC}$:

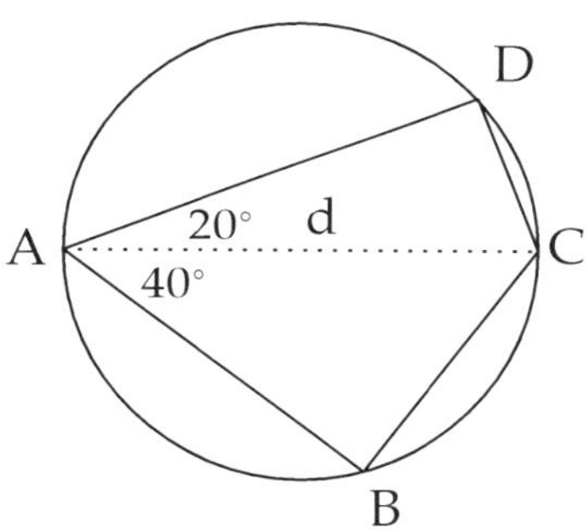

Let $AC = d$. Then $AB = d\cos 40°$, $BC = d\sin 40°$, $CD = d\sin 20°$, and $DA = d\cos 20°$. By Ptolemy's theorem:

$$(d\cos 40°)(d\sin 20°) + (d\sin 40°)(d\cos 20°) = 6d$$

However the left-hand side simplifies easily to

$$d^2(\cos 40° \sin 20° + \sin 40° \cos 20°) = d^2 \sin 60° = \frac{\sqrt{3}d^2}{2}$$

by the sine addition formula. Therefore $\frac{\sqrt{3}d^2}{2} = 6d \implies d = \boxed{4\sqrt{3}}$. □

21. We have $\log_{2^x}(\log_{2^y} 2^{400}) = 1 \implies \log_{2^y} 2^{400} = 2^x$, or $(2^y)^{2^x} = 2^{400}$. Equating the exponents, we see that $y2^x = 400$. We want to maximize $x + y$, so set $y = 200$ and $x = 1$. This is maximum since y and 2^x are positive integer divisors of 400. Then $x+y = \boxed{201}$. □

22. We do casework based on the location of the 1.

If the 1 is in the center cell, then the 2 can be in any of the four cells adjacent to 1, and the 3 can be in either of the two cells adjacent to the 2. This uniquely determines the rest of the grid, giving $4 \times 2 = 8$ ways.

If the 1 is in any of the four corner cells, there are 4 ways to choose the location of the 1. Without loss of generality, suppose the 1 is in the upper left cell. There are two possible locations for the 2; suppose the 2 is in the upper cell, as shown:

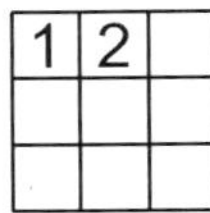

The 3 must go in either the top right cell or the center cell. We enumerate all possibilities:

1	2	3
8	7	4
9	6	5

1	2	3
8	9	4
7	6	5

1	2	3
6	5	4
7	8	9

1	2	9
4	3	8
5	6	7

Hence for any of the $4 \times 2 = 8$ ways to insert 1 and 2 (where 1 is in a corner cell), there are 4 possible grids, giving $4 \times 2 \times 4 = 32$ grids.

If the 1 is in any of the four "side" cells (any cell which is not the center or a corner), it can be seen that there is no such grid satisfying the requirements.

Therefore the number of possible grids is $8 + 32 = \boxed{40}$. □

23. The main insight is that any set of three lines for which no two are parallel, and which do not all intersect at the same point, determine one and only one triangle. Hence we can count the number of sets of three lines that satisfy these properties.

Assume hexagon $P_1P_2P_3P_4P_5P_6$ is drawn in the standard xy-plane such that P_1P_2 is horizontal. Out of the $\binom{6}{2} = 15$ lines, three lines have slope zero ($\overleftrightarrow{P_1P_2}$, $\overleftrightarrow{P_3P_6}$, and $\overleftrightarrow{P_4P_5}$), three lines have slope $\sqrt{3}$, three lines have slope $-\sqrt{3}$, two lines have slope $\frac{\sqrt{3}}{3}$, two lines have slope $-\frac{\sqrt{3}}{3}$, and two lines have undefined slope (vertical). Note that the actual slopes are unimportant, only the number of pairs of parallel lines. Hence, the 15 lines can be partitioned into sets of parallel lines with sizes 3, 3, 3, 2, 2, 2.

We want to find the number of ways to choose three of these 15 lines with no two lines parallel. We use some casework:

Case 1: All three lines are chosen from the sets of size 3. This gives $3^3 = 27$ ways of choosing three nonparallel lines.

Case 2: Two lines are chosen from sets of size 3, and one line is chosen from a set of size 2. This gives $\binom{3}{2} \times 3^2 \times \binom{3}{1} \times 2 = 162$ ways.

Case 3: One line is chosen from a set of size 3, and two lines are chosen from sets of size 2. This gives $\binom{3}{1} \times 3 \times \binom{3}{2} \times 2^2 = 108$ ways.

Case 4: All three lines are chosen from the sets of size 2. This gives $2^3 = 8$ ways.

Altogether we get $27 + 162 + 108 + 8 = 305$ ways to choose 3 lines for which no two are parallel (alternatively, we can complementary count by subtracting the number of ways to choose 3 lines for which some two are parallel, then subtract from $\binom{15}{3} = 455$).

However we must subtract the number of sets of three lines which intersect at the same point. With some careful drawing/inspection, we see that the only ways for three lines $\overleftrightarrow{P_iP_j}$ to intersect at the same point is if they intersect at one of the vertices of the hexagon, or at the center. If the three lines intersect at a vertex (say P_1), there are $\binom{5}{3} = 10$ ways to choose three lines that intersect at P_1, then 6 ways to choose the vertex which serves as the common intersection, giving 60 ways. If the three lines intersect at the center of the hexagon, there is only 1 way to choose 3 lines which intersect at the center ($\overleftrightarrow{P_1P_4}$, $\overleftrightarrow{P_2P_5}$, and $\overleftrightarrow{P_3P_6}$).

Altogether we get $305 - 60 - 1 = 244$ ways to choose three lines which are pairwise not parallel and do not intersect at a common point; each way determines a unique triangle, so the answer is $\boxed{244}$. □

24. Since $\angle EBC = 90°$, $\overline{EC}$ must be a diameter of ω, as in the following diagram:

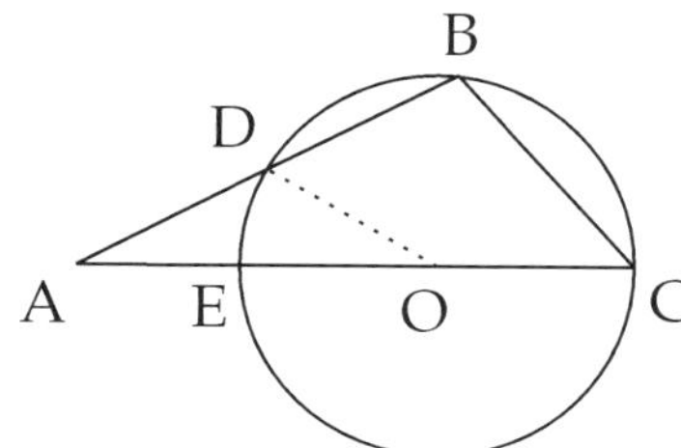

Without loss of generality, suppose the radius of ω is 1, so that $AE = 1$ and $EC = 2$. By Power of a Point about A, we have $1 \cdot 3 = AD \cdot AB = 2AD^2$, so $AD^2 = \frac{3}{2} \implies AD = DB = \frac{\sqrt{6}}{2}$.

Let $BC = x$ so that $BE = \sqrt{4 - x^2}$, and let O be the center of circle ω. We know that $\triangle ODA$ has side lengths 1, 2, and $\frac{\sqrt{6}}{2}$. By law of cosines on $\triangle ODA$:

$$1 = 4 + \frac{3}{2} - 2\sqrt{6}\cos\angle A$$

We can solve for $\cos\angle A$, giving $\cos\angle A = \frac{9}{4\sqrt{6}}$. Applying the law of cosines on $\triangle ABC$:

$$\begin{aligned} x^2 &= 9 + 6 - 6\sqrt{6}\cos A \\ &= 15 - 6\sqrt{6}\left(\frac{9}{4\sqrt{6}}\right) \\ &= \frac{3}{2} \end{aligned}$$

Then $BC^2 = \frac{3}{2}$, and by the Pythagorean theorem $BE^2 = 4 - \frac{3}{2} = \frac{5}{2}$. The desired ratio is $\frac{\frac{5}{2}}{\frac{3}{2}} = \boxed{\frac{5}{3}}$. □

25. Let E_0 denote the desired answer, and for $i = 1, 2$, let E_i denote the expected number of dice rolls needed, given that the previous i consecutive rolls were the same. We set up the following system of equations:

$$E_0 = 1 + E_1 \tag{3}$$
$$E_1 = \frac{5}{6}(1 + E_1) + \frac{1}{6}(1 + E_2) = 1 + \frac{5}{6}E_1 + \frac{1}{6}E_2 \tag{4}$$
$$E_2 = \frac{5}{6}(1 + E_1) + \frac{1}{6}(1) = 1 + \frac{5}{6}E_1 \tag{5}$$

We solve for E_1 and E_2. Using (5), we can substitute E_2 with $1 + \frac{5}{6}E_1$ into Eq. (4) to obtain

$$E_1 = E_2 + \frac{1}{6}E_2 = \frac{7}{6}E_2$$

or equivalently $E_2 = \frac{6}{7}E_1$. Substitute E_2 back into Eq. (4):

$$E_1 = 1 + \frac{5}{6}E_1 + \frac{1}{6}\left(\frac{6}{7}E_1\right)$$

Solving for E_1, we get $E_1 = 42$. The desired answer is $E_0 = 1 + E_1 = \boxed{43}$. □

Made in the USA
Coppell, TX
21 January 2023

11444990R00105